U.S. Stabling Guide

Eleventh Edition

LARIAT PUBLICATIONS
P.O. Box 364, Kingston, Massachusetts 02364

1-800-829-0715
1-781-829-0710
www.usstablingguide.com

U.S. Stabling Guide

The Country's Comprehensive Guide for Horse Transportation

P.O. Box 364, Kingston, MA 02364
800-829-0715

Publisher: Lisa A. Doubleday

Editor: Jean Cohen

Graphic Artist: Cherry Bishop, Graphics Plus
Duxbury, Massachusetts

Digital Photography: Gabriel Lortie
Middleboro, MA
508-789-2654
www.gabriellortie.com
(cover photography)

PUBLISHER'S NOTICE
Every effort is made to compile and print this directory as accurately as possible. However, errors or omissions may occur. In the event of an error or omission, Lariat Publications assumes no obligation to correct the same in this directory or by any special notice of any kind to the individuals and businesses listed herein.

Copyright © 2005 by Lariat Publications. Printed in the U.S.A.

All rights reserved. No part of this work may be reproduced or transmitted in any form or by any means, electronic or mechanical, including photocopying and recording, or by any information storage or retrieval system, except as may be expressly permitted by the 1976 Copyright Act or in writing from the publisher.

TABLE OF CONTENTS

Advertisers Directory	9
How to Use the Guide	11
Feeding the Traveling Horse	13
Horse Transportation Companies	17
State Transportation Requirements	18
Hat Etiquette	20
ALABAMA	22
ARIZONA	26
ARKANSAS	34
CALIFORNIA	38
COLORADO	44
CONNECTICUT	54
DELAWARE	56
FLORIDA	58
GEORGIA	64
IDAHO	68
ILLINOIS	72
INDIANA	78
IOWA	82
KANSAS	84
KENTUCKY	88
LOUISIANA	90
MAINE	94
MARYLAND	98
MASSACHUSETTS	102
MICHIGAN	108
MINNESOTA	112

TABLE OF CONTENTS

MISSISSIPPI	116
MISSOURI	120
MONTANA	128
NEBRASKA	136
NEVADA	140
NEW HAMPSHIRE	144
NEW JERSEY	148
NEW MEXICO	152
NEW YORK	158
NORTH CAROLINA	164
NORTH DAKOTA	168
OHIO	172
OKLAHOMA	178
OREGON	184
PENNSYLVANIA	190
RHODE ISLAND	196
SOUTH CAROLINA	198
SOUTH DAKOTA	202
TENNESSEE	206
TEXAS	216
UTAH	226
VERMONT	230
VIRGINIA	234
WASHINGTON	242
WEST VIRGINIA	246
WISCONSIN	248
WYOMING	252
ALBERTA, CANADA	261

TABLE OF CONTENTS   PAGE 7

BRITISH COLUMBIA, CANADA	261
MANITOBA, CANADA	262
NEW BRUNSWICK, CANADA	262
NOVA SCOTIA, CANADA	263
ONTARIO, CANADA	263
SASKATCHEWAN, CANADA	265
About the Publisher	277

courtesy of

TRAILERING TIPS

Equine Precautions – Short Trip

Anytime a horse is loaded into a trailer, whether for a short trip or a long one, these fundamental measures should be taken:

- **Train your horse to load calmly and to accept the trailer as non-threatening.** The best defense against injury and illness is good training.

- **Make sure your trailer is safe.** Once a horse has been trained to trust you and the trailer, don't let it let them down.

- **Drive carefully.** Remember you have live cargo in the trailer – drive accordingly.

- **Make sure all inoculations are current.** Current inoculations will protect your horse from exposure to other horses.

- **Wrap all four legs.** Just walking into a trailer can result in injury if the horses scrapes against something, so wrap legs every time your horse gets on the trailer.

- **Make sure trailer is vented.** Horse are very sensitive to dust and noxious gasses; i.e., such as ammonia from urine and manure.

- **Carry an emergency first aid kit.** Keep it in your trailer and make sure it is always ready and up-to-date.

- **Learn proper first aid techniques.** Learn how to bandage wounds in various locations, control blood loss, and learn to recognize the signs of dehydration, heat exhaustion, and colic.

- **Learn how to monitor vital signs in the horse.** If your horse is sick or hurt, you can give the veterinarian current vital signs via telephone.

- **Carry backup supplies appropriate to the length of the trip.** Keep in mind your trip may be longer than planned due to unforeseen circumstances.

- **Carry a medical ID.** If you are incapacitated in an accident, it can be important to contact someone who knows you and your horses.

Equine Trailering Tips courtesy of USRider Equestrian Motor Plan in cooperation with nationally respected equine trailering expert and author Neva Kittrell Scheve and veterinarian and author James M. Hamilton, DVM. For more trailering information, visit the Equine Trailer Safety Area at www.usrider.org.

www.usrider.org

ADVERTISERS DIRECTORY

COMPANY	PAGE
US Rider	1
US Rider Trailering Tips	throughout guide
Bar "B" Farms	181
Black Forest Farm, LLC	249
Blue Hank Horse Stable	171
Brian's Trailer Sales	81
Caution Horses	inside front cover
Cruse Acres	224
DJ Bar Ranch	129
Double D Ranch	201
DRA Dude Ranch Association	288
Equestrians Edge Book Club	12
Graphics Plus	269
Hidden Lakes Stable	225
Horsein' Around	269
Meacham Ranch	143
Mohave County Fairgrounds	29
Our Hands for Horses	104
Ride Pennsylvania Horse Trails	195
Shelby Farms Showplace	211
Valar Horse Boarding & Transport	269
www.USStablingGuide.Com	inside back cover

courtesy of

TRAILERING TIPS

Equine Precautions – Long Trips

The importance of these precautions is directly proportional to the length of the trip. Use your own judgment — **and that of your veterinarian.**

- **Electrolytes** – Increase 2 to 3 days prior to shipping.
- **Bran Mash** – Once a day for 2 to 3 days prior to shipping.
- **Vitamins** – Add extra for a week prior to shipping.
- **Mineral Oil** – One pint per day may either be added to feed along with bran for four days prior OR given by veterinarian via stomach tube the day of shipping (4-6 hours before departure).
- **Antibiotics** – When the trip will be longer than 12 hours, discuss the administration of antibiotics with your veterinarian.
- **Body Clip** – When taking your horse from a cold climate to a warm one, a body clip is recommended.
- **Blanket** – The need for the blanket will depend on the temperature en route.

After Prolonged Storage — Inspection Procedures

Before removing trailer from jack stands:

1. **Remove all wheels and hubs or brake drums.** (Note which spindle and brake the drum was removed from so it can be reinstalled in the same location).
2. **Inspect suspension for wear.**
3. **Check tightness** of hanger bolt, shackle bolt, and U-bolt nuts per recommended torque values.
4. **Check brake linings, brake drums, and armature faces** for excessive wear or scoring.
5. **Check brake magnets with an ohmmeter.** The magnets should check 3.2 ohms. If shorted or worn excessively, replace.
6. **Lubricate all moving parts,** using a high-temperature brake lubricant (Lubriplate or the equivalent). *Caution: Do not get grease or oil on brake linings or on magnet face.*
7. **Remove any rust from braking surface and armature surface of drums** with fine emery paper or crocus cloth. Protect bearings from contamination while removing rust.
8. **Inspect oil or grease seals for wear or nicks.** Replace if necessary.

Equine Trailering Tips courtesy of USRider Equestrian Motor Plan in cooperation with nationally respected equine trailering expert and author Neva Kittrell Scheve and veterinarian and author James M. Hamilton, DVM. For more trailering information, visit the Equine Trailer Safety Area at www.usrider.org.

www.usrider.org

HOW TO USE THE GUIDE

1. States are listed alphabetically with cities and towns within the state also listed alphabetically. Canadian provinces follow Wyoming and are also listed alphabetically.

2. Accompanying state maps provide quick visual locations of stables and major highways but should not be used as your map while you are traveling. Use accurate road maps for that purpose.

3. Each page has a written reminder that all stables require current health papers. Some stables require proof of additional vaccinations, and that is included in their listing. Our advice to all of the listing stables is that they not allow the unloading of horses until those papers are produced. Please discuss all of the traveling medical requirements with your veterinarian well before you begin your trip.

4. Stable listings provide the owner and/or manager's name, telephone number(s), address, a description of facilities, current overnight rates per horse, and, in many cases, concise directions to the stable. Unless otherwise stated, do not drive directly to a stable without prior notification of your arrival. Preparations usually have to be made before overnight boarders arrive.

5. "Accommodations" lists the nearest motels to the stable for your own overnight stay. This has been included so you can make your reservations at a motel or bed & breakfast and be assured of a good night's rest before beginning another day on the road.

6. For your interest, we have included additional information about breeding programs, unique services or products available, local activities or sights to see while you are in a new area, etc.

7. We are not a reservation service and request that you deal directly with the stables in making your overnight arrangements. However, we would be happy to answer any general questions about traveling with your horse.

8. We have made every effort we can at this time to assure that the stables in our book will provide a clean and safe environment for your horse. However, it is impossible to inspect each stable. If you feel that any particular stable fails to provide a clean and safe environment, notify us. After receiving two or more similar complaints, we will arrange for an inspection of that stable.

We Are Proud to Announce Our Book:

The U.S. Stabling Guide

Is Now Available Through
The Equestrian's Edge Book Club.

FEEDING THE TRAVELING HORSE
By Tara Devine, P.C.H.A.

In order to make your horse's journey comfortable, you need to look at certain factors that will affect the horse during the trip. First, how far are you traveling with the horse? Is the trip going to involve 8 hours of traveling or 8 days? A long trip will require more planning and caution when feeding your horse than a short trip will. Second, what are the weather conditions going to be like? Will it be 90 degrees and humid or 30 degrees and raining? Intense heat and humidity can really add to the stress of travel, just as extreme cold temperatures will require the horse to create more body heat. Third, how do you expect the horse to handle the stress of traveling? Is the horse a seasoned campaigner who is used to long hauls or is the horse a nervous Nelly who goes off feed when upset? Your answers to these questions are clues that will help you care for your horse so he/she will be comfortable and in good condition when you arrive at your destination.

Providing enough water is a crucial part of keeping the horse healthy during the ride. Water should be offered every two to three hours. If possible, try to carry some of your horse's current water supply with you in spill-proof containers (new, clean gas jugs work well). Many horses may refuse to drink different water because it smells or tastes funny to them. If this happens, you can add small amounts of flavoring to the new water (try molasses or powdered drink mixes). The horse may also be more comfortable if you bring his/her old familiar water bucket along. Be sure to monitor the amount of water that your horse drinks. Compare the quantity to what he/she normally drinks at home. Generally speaking, a 1000-lb. idle horse will drink about 8-10 gallons of water a day. Realize that if the weather is very hot and humid the horse will have a higher water requirement in order to replace the water that is lost during sweating. If you are concerned that the horse is not drinking enough water, you can help by giving a sloppy bran mash or by adding water to the regular ration of grain. Water is crucial to a horse's digestion and ability to regulate body temperature. It is probably the most critical factor in keeping your horse in good health during a trip.

One of the biggest components of feeding a horse on the road (or at home) is offering hay. In general, you should provide hay in a hay net at all times during the trip. By munching on hay constantly the horse will keep essential bacteria in the gut alive and healthy. These bacteria are necessary for the proper digestion of food. Providing free choice hay will help keep the horse occupied and will also provide the horse with a substantial portion of the calories and nutrients he/she needs for the day. It is best if you bring your own supply of hay with you so that the horse's digestive system is not subjected to different hay. If you cannot bring hay along, then be sure to buy bales along the way that are similar to what you were feeding. For example, try not to go from a local grass hay to a straight alfalfa hay. It is too drastic a change. If you are concerned that the new hay will be very different, save enough of your old hay to blend with the new. This will help your horse gradually become accustomed to the new hay.

Whether or not you feed grain during your trip is really dependent on the questions presented in the first paragraph. During a short trip when the horse would only be missing a meal or two, you would probably be better off not to feed any grain. The hay that you are offering will provide them with the calories and nutrients he/she needs for the day. But if it is a longer trip, it may be necessary to feed the horse some of his/her normal ration. How much depends on what the normal ration is based on. If you normally ride your horse a couple of hours a day, his/her grain ration contains enough calories to support that amount of work. But once the horse is on the trailer, he/she will not be burning as many calories so you must cut back on the feed. Also take a look at the horse's personality and how he/she is reacting to the trip. Nervous horses present a problem because they tend to need the extra calories that the grain provides. On the other hand, they are more prone to colic and other digestive upsets. If your horse seems to be extremely agitated, you would be better off not to feed grain and just offer hay. If you do feed grain, offer it in small feedings at least three times a day. Smaller portions offered more frequently are the best way to go to avoid stomach upsets. Be sure to bring your grain with you or use a commercial feed like Purina that is available throughout the country. It is critical not to switch your horse's grain during the trip (even more so than the hay) to avoid serious digestive problems. If you know that you will have to switch feeds during the trip, bring enough of the old feed along so you can gradually blend in the new feed.

By planning ahead and observing your horse's reaction to the stress of traveling, you will be able to feed your horse so he/she will arrive healthy and happy. Happy trails!

[Publisher's note: Tara Devine is a Purina Certified Horse Advisor. She also has a B.A. degree in Management, has been a horse nutritionist for ten years, conducts horse nutrition seminars, and owns a feed and tack store. Tara has been a horse owner all of her life.]

courtesy of
USRider

TRAILERING TIPS

Trip Preparation Checklist for Trailer Axle Assembly

Taking care of your trailer axle assembly can add to its life—and may protect your own life as well. Using the following checklist before starting a trip with your trailer is highly recommended. Some of these items should be checked 2 to 3 weeks before a planned trip to allow sufficient time to perform maintenance.

- ☐ Check your maintenance schedule and be sure you are up to date.
- ☐ Check hitch. Is it showing wear? Is it properly lubricated?
- ☐ Fasten safety chains and breakaway switch actuating chain securely. Make certain the breakaway battery is fully charged.
- ☐ Inspect towing hookup for secure attachment.
- ☐ Load your tag-a-long trailer so that approximately 10 percent of the trailer's total weight is on the hitch. For light trailers this should be increased to 15 percent.
- ☐ Do not overload. Stay within your gross vehicle rated capacity.
- ☐ Inflate tires according to manufacturer's specifications; inspect tires for cuts, excessive wear, etc.
- ☐ Check wheel mounting nuts and bolts with a torque wrench. Torque in proper sequence to specified values.
- ☐ Make certain of hanger bolt, shackle bolt, and U-bolt nuts per torque values specified.
- ☐ Make certain that brakes are synchronized and functioning properly.
- ☐ Check operation of all lights.

Equine Trailering Tips courtesy of USRider Equestrian Motor Plan in cooperation with nationally respected equine trailering expert and author Neva Kittrell Scheve and veterinarian and author James M. Hamilton, DVM. For more trailering information, visit the Equine Trailer Safety Area at www.usrider.org.

www.usrider.org

courtesy of
USRider

TRAILERING TIPS
Don't Go on the Road Without It!

For the Trailer

Store these items in the trailer so you always have them on hand.

- **Spare tire, Jack, Tire iron**
- **Three emergency triangles or flares** (Triangles are preferred)
- **Wheel Chock**
- **Flashlight**
- **Electrical tape**
- **Duct Tape**
- **Equine First Aid Kit with splint** (know how to use it.)
- **Knife for cutting ropes, etc., in emergency**
- **Water**
- **Buckets, Sponge**
- **Water hose**
- **Spare halter and lead rope for each horse**
- **Spare bulbs for exterior and interior lights**
- **Spare fuses if applicable**
- **Fire extinguisher** (with up-to-date charge)
- **WD-40 or other lubricant**
- **Broom, Shovel, Fork and Manure disposal bags**
- **Insect spray** (bee and wasp)

For the Tow Vehicle:

- **Registration for the vehicle and trailer**
- **Proof of insurance**
- **Jumper cables**
- **Tool kit including wiring materials**
- **Spare belts and hoses for the tow vehicle**
- **Tow chain**
- **Work gloves**
- **Cellular phone and/or CB radio** (CB may be more effective in rural areas)
- **Replacement fuses**
- **Portable air compressor**
- **Extra cash/credit card**
- **Road atlas**
- **Hawkins Guide:** Equine Emergencies on the Road
- **Active USRider Membership**

- Check your inventory frequently and replace used or removed items before each trip.
- For crossing state lines or attending competitions:
  Certificate of Veterinary Inspection (Health Certificate) dated within 30 days.
  Proof of Negative EIA (Coggins) usually dated within 1 year. Some states require within 6 months.
  Certificate of Brand Inspection if applicable.
- **Emergency Directions.** If you are in an accident and have been injured yourself – EMS personnel and police will most likely not be capable of taking care of your horses. Prepare for this situation by keeping some sort of emergency directions in a very visible place. Write the name and current telephone number(s) of someone you know who can be called to help or to advise what to do with the horses if you are incapacitated, such as — a knowledgeable friend, your veterinarian, or someone else who is familiar with your horses. **The USRider Membership Kit** contains a specially created emergency information placard (trailer interior) and an accompanying emergency notification sticker (trailer exterior).

Equine Trailering Tips courtesy of USRider Equestrian Motor Plan in cooperation with nationally respected equine trailering expert and author Neva Kittrell Scheve and veterinarian and author James M. Hamilton, DVM. For more trailering information, visit the Equine Trailer Safety Area at www.usrider.org.

www.usrider.org

HORSE TRANSPORTATION COMPANIES

The following is a list of individuals or companies that offer local, regional, and/or national transportation for one or more horses. Please refer to the page number following their names and addresses where further information about their services can be found.

NAME	PAGE
AJ Stables - Westfield, Massachusetts	106
CR Livestock & Animal Care - Longmont, Colorado	49
Cumberland Springs Ranch - Knoxville, Tennessee	209
Diamond T Transportation - Willis, Texas	223
Eastern Equine Express - Florence, South Carolina	199
Gee Jay Ranch - Cherry Valley, California	40
Horsein' Around	269
Lewis Stables - Slidell, Louisiana	92
MNMS Stables - Plainfield, New Hampshire	146
Terry Teeples - South Jordan, Utah	228
Valar Horse Boarding & Transport - Jackson, Wyoming	269

STATE TRANSPORTATION REQUIREMENTS

Below is basic information on the health requirements of the 50 states for transient equine. For detailed information or questions, please call that state's veterinarian's office. "E.I.A." is Equine Infectious Anemia and "Within" is number of months within which the E.I.A. test must have been done. "Hlth. Cert." is whether or not a health certificate is required and "Special" is any other tests, information, etc. needed.

STATE	E.I.A.	WITHIN	HLTH.CERT.	SPECIAL
ALABAMA	yes	12 mos.	yes	None
ALASKA	yes	6 mos.	yes	H.C. to state no ectoparasites
ARIZONA	no	N/A	yes	H.C. to include brands &/or tattoos
ARKANSAS	yes	12 mos.	yes	Temp. reading & date/lab on H.C.
CALIFORNIA	yes	6 mos.	yes	Note on H.C. if horse is transient
COLORADO	yes	12 mos.	yes	Permit if E.I.A. pending
CONN.	yes	12 mos.	yes	Temp. req. on H.C.
DELAWARE	yes	12 mos.	yes	Temp. reading req.
FLORIDA	yes	12 mos.	yes	LV Rhinopneumonitis vacc. not eligible for 21 days after vacc. Temp. reading on H.C.
GEORGIA	yes	12 mos.	yes	Equidae (6 mos.) not w/dam - test
HAWAII	yes	3 mos.	yes	Call 808-487-5351
IDAHO	no	N/A	yes	None
ILLINOIS	yes	12 mos.	yes	Check w/ race track if that is destination
INDIANA	yes	12 mos.	yes	Foals w/dam exempt from test
IOWA	yes	12 mos.	yes	E.I.A. date/lab on H.C.
KANSAS	no	N/A	yes	None
KENTUCKY	yes	6/12 mos.	yes	Horses for sale tested w/in 6 mos., others w/in 12
LOUISIANA	yes	12 mos.	yes	E.I.A. test info on H.C.
MAINE	yes	6 mos.	yes	None
MARYLAND	yes	12 mos.	yes	Temp. reading on H.C.
MASS.	yes	6 mos.	yes	Name of lab req.
MICHIGAN	yes	6 mos.	yes	Preapproved H.C. for exhibitions
MINNESOTA	yes	12 mos.	yes	Permit if E.I.A. pending. Suckling foals exempt

STATE TRANSPORTATION REQUIREMENTS
(continued)

STATE	E.I.A.	WITHIN	HLTH.CERT.	SPECIAL
MISSISSIPPI	yes	12 mos.	yes	Orig. test cert. req.
MISSOURI	yes	6 mos.	yes	Call 314-751-3377 for info. on VEE vacc.
MONTANA	no	6 mos.	yes	Prior approval of H.C. req.
NEBRASKA	yes	12 mos.	yes	Name of lab on H.C.
NEVADA	no	6 mos.	yes	Neg. E.I.A. & permit req. if going to rodeo finals
N. HAMPSHIRE	yes	6 mos.	yes	None
NEW JERSEY	yes	12 mos.	yes	E.I.A. test date/lab on H.C.
NEW MEXICO	yes	12 mos.	yes	E.I.A. test date/lab on H.C.
NEW YORK	yes	12 mos.	yes	E.V.A. test w/in 30 days on TB stallions req.
N. CAROLINA	yes	12 mos.	yes	None
N. DAKOTA	yes	12 mos.	yes	Suckling foals exempt; AZ horses exempt from E.I.A. test
OHIO	yes	6 mos.	yes	Temp. reading on H.C.
OKLAHOMA	yes	6 mos.	yes	E.I.A. test date/lab info. on H.C.
OREGON	yes	6 mos.	yes	Permit: 503-378-4710
PENN.	yes	12 mos.	yes	Suckling foal exempt
R. I.	no	N/A	yes	Prior approval of H.C. 401-277-2781
S. CAROLINA	yes	12 mos.	yes	None
S. DAKOTA	yes	12 mos.	yes	H.C. within 10 days
TENNESSEE	yes	6 mos.	yes	None
TEXAS	yes	12 mos.	yes	Permit req. on slaughter equine. 512-479-6697
UTAH	yes	12 mos.	yes	Suckling foals exempt from test
VERMONT	yes	12 mos.	yes	Permit 802-828-2450
VIRGINIA	yes	12 mos.	yes	None
WASH.	yes	6 mos.	yes	None
W. VIRGINIA	yes	12 mos.	yes	Suckling foals exempt
WISCONSIN	yes	12 mos.	yes	Suckling foals exempt All E.I.A. test info on H.C
WYOMING	yes	12 mos.	yes	All E.I.A. test info. on H.C.

Hat Etiquette (1880)

A good cowboy hat will bring many years of pleasure, especially if you follow these simple rules of hat care and etiquette:

Never handle your hat by the crown. This breaks the fibers of the fur felt. Always pick up a hat by the brim.

When handing another man his hat, hand it to him crown-down with the back facing you; this way, he can simply flip it on his head without having to turn it around.

Never park a hat in a hot car or on a radiator. Heat will make it shrink.

Always store a hat upside-down so the roll of the brim won't flatten out.

Never put a hat on a bed or bunk; that's bad luck. Cowboys customarily stored their hats on the bunkhouse floor right beside their bed.

Don't try on another man's hat. This is a serious breach of etiquette, almost as bad as getting on his horse.

Identify your hat inside your brim. Most good hat shops give you a free insert that says, "Like hell it's yours. This hat belongs to _____."

— "Victorian Etiquette, Social Graces and Manners" by Cherly Miller

U.S. Stabling Guide

State Listings

Eleventh Edition

LARIAT PUBLICATIONS
P.O. Box 364, Kingston, Massachusetts 02364

1-800-829-0715
1-781-829-0710
www.usstablingguide.com

Page 22 ALABAMA

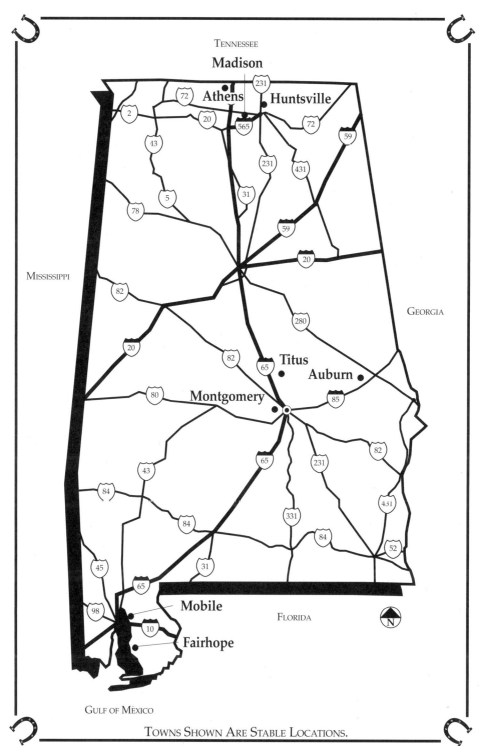

Towns Shown Are Stable Locations.

∗ onsite accommodations

ALABAMA PAGE 23

ALL OF OUR STABLES REQUIRE CURRENT NEG, COGGINS, CURRENT HEALTH PAPERS & OWNERSHIP PAPERS.

ATHENS
Red Pepper Ranch   Phone: 256-233-7174
Joanne R. Fellows   E-mail: fellcamp@bellsouth.net
26042 Pepper Rd. [35613] **Directions:** Call for directions. **Facilities:** 2 12x12 stalls, 2-acre turnout with run in shed. Vet within 3 miles. **Rates:** $15 per night. **Accommodations:** B&B soon to be available.

AUBURN
Diamond D Stables   Phone: 334-821-6664
Richard Garcia or Peggy Durell
610 Lee Road 395 [36830] **Directions:** From I-85: Take Exit 51, Auburn University. Go south on Hwy 29 to mile marker #177. Follow signs to "Diamond D." **Facilities:** 5 indoor stalls, pasture/turnout; 10 minutes from large animal clinic. **Rates:** $25 per day, $100 per week. **Accommodations:** Auburn University Hampton Inn 2 miles from stable.

FAIRHOPE
Lakewood Stables   Phone: 251-945-6711
Sharon & Bill Adams
12960 County Rd. 48, Silver Hill [36576] **Directions:** I-10 to 98 South to Fairhope. Call for directions. **Facilities:** 44 indoor stalls, 30 acres fenced pasture/turnout, feed/hay, trailer parking. Reservations required. Call by 6:00 P.M. **Rates:** $15 per day; $75 per week with advance notice. **Accommodations:** Marriot Grand Hotel, 2 miles from stable; Baron's Motel, 3 miles; Holiday Inn Express, approx. 3 miles.

HUNTSVILLE
Flint Ridge Farm, Inc.   Phone: 256-776-3635
Robert & Diana Rose
3616 Maysville Road [35811] **Directions:** Approx. 4 miles from Hwy 72 off of I-65. Call for directions. **Facilities:** 2 indoor stalls, feed/hay available, trailer parking, trail riding, hunter/jumper, dressage instruction. No smoking, no alcohol, reservations required. **Rates:** $25 per day, weekly available with advance notice. **Accommodations:** Comfort Inn 15 minutes from stable.

MADISON
Rainbow Riding Academy, Inc.   Phone: 256-830-2911
Patricia Whitfield   or: 256-837-7758
212 Capshaw Road [35757] **Directions:** Approx. 15 miles from I-65. Take I-65 to I-565 East to Rideout Road North to Hwy 72 West. Right on Jeff Road, left at first traffic light onto Capshaw Road. **Facilities:** 6-10 indoor 10' x 16' stalls, 4 outside paddocks, 40 acres pasture/turnout, wash stall (hot and cold water), outside working pen, walker, lighted outside arena, hay/feed available, trailer parking. 10 minutes to several vets. Western lessons, parties, hay rides; carousel with ponies for rent. **Rates:** $20 per day. **Accommodations:** Close to numerous motels/hotels, restaurants, and shopping centers. 10 minutes to space and rocket center.

ALABAMA

ALL OF OUR STABLES REQUIRE CURRENT NEG. COGGINS, CURRENT HEALTH PAPERS & OWNERSHIP PAPERS.

MONTGOMERY
Double D Stables   Phone: 334-284-4982
Benjamin & Connie D'Amico
124 Mizell Drive [36116] **Directions:** Call for directions. **Facilities:** 4 - 12' x 12' stalls, 12' x 24' also available, turnout runs with stalls, 1 acre pasture per horse, trailer parking, secured premises, feed/hay, vet within 10 minutes of facility. **Rates:** 12' x 12': $15 per day, $100 per week; 12' x 24': $35 per day. **Accommodations:** Econo Lodge within 6 miles.

Harness Hill Stables   Phone: 334-272-9059
Keith Valley
#2 Harness Hill Drive [36116] **Directions:** Taylor Road South exit off of I-85. Call for directions. **Facilities:** 16 indoor 12' x 12' box stalls, show barn, outdoor covered arena, covered round pen, 14' x 14' stallion stall, sky lights in every stall, ventilation fans, and natural gas heat in barn. Stable is on 150 acres and has Western tack & saddle shop on premises. Training of horses for Western pleasure and dressage. Call for reservations. **Rates:** $15 per day; ask for weekly rate. **Accommodations:** Mariott 4 miles from stable.

TITUS
Lucky 7 Ranch   Phone: 334-567-9752
Sherry Moore   E-mail: CactusRacer@aol.com
Web: www.geocities.com/luck7barrels/index.html
180 Grass Farm Road [36080] **Directions:** From I-65: Take Hwy 14 East from prattville to US 231 North. Go approximately 11 miles to Elmore cty 80, turn right, 3rd Driveway on right. From I-85- Take exit 6 in Montgomery and turn right onto Hwy 231. Go north 25 miles. Turn right on Elmore cty 80. 3rd driveway on right. **Facilities:** 4 10x12 indoor stalls, trailer parking available, 7 acres of pasture, round pen, 150x120 arena, Please call 24 hours in advance. **Rates:** $20 per stall, $15 per pasture or outside pen. **Accommodations:** Key West Inn-Wetumpka, AL (15 miles).

NOTES AND REMINDERS

ARIZONA

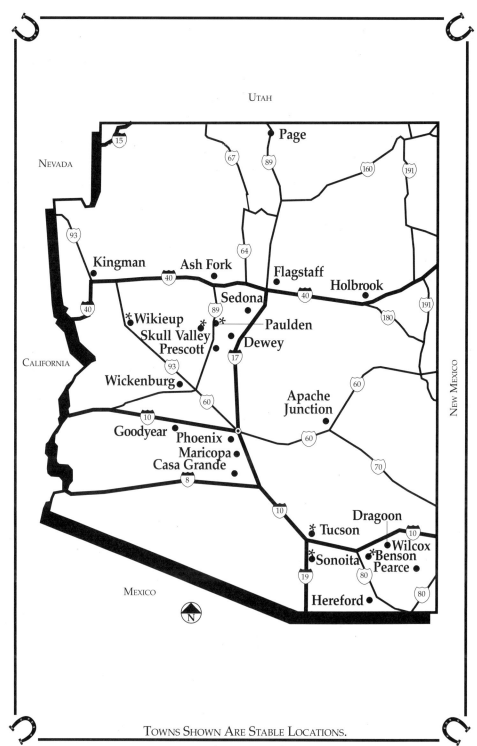

Towns Shown Are Stable Locations.

* onsite accommodations

ARIZONA PAGE 27

ALL OF OUR STABLES REQUIRE CURRENT NEG. COGGINS,
CURRENT HEALTH PAPERS & OWNERSHIP PAPERS.

APACHE JUNCTION
<u>OK Corral Stables</u>　　　　　　　　　　　　　Phone: 480-982-4040
Ron & Jayne Feldman　　　　　　　　　　　Web: www.okcorrals.com
P.O. Box 528 [85217] Directions: Highway 60 to N. Tomahawk Exit. North on Tomahawk 5 miles. **Facilities:** 25 - 12' x 24' corrals, 70" round pen, miles of riding trails nearby, easy access to Superstition Mountain Wilderness. Feed/hay, horses bought, sold, & rented. No stallions overnight. Call for reservations. **Rates:** $15 per day; $40 per week; $75 monthly. **Accommodations:** RV hookups. Holiday Inn, Super 8 and many motels within 3-4 miles.

<u>Quarter Circle Boot Ranch</u>　　　　　　　　Phone: 480-983-7270
Marnie Young
1276 N. Wickiup Rd(85219) Directions: 2 1/2 miles north of hwy 60 on Tomahawk to Superstition, east 2 blocks to Wickiup, north 1/4 mile. **Facilities:** 32 horse facility, 20 20x20 pens, 8 16x16 pens, 5 box stalls with runs. 140'x 240' arena, 40' round pen. Alfalfa cubes and pellets available. Parking for horse trailers only. 4 horse hot walkers, 2 wash racks, close to trails in Superstition and Goldfield Mtns. Veterinary health certificates and coggins. **Rates:** $10 w/o feed $15 w/feed, $130 per month. **Accommodations:** RV Park directly across the street, many other RV Parks within a mile, motels within 2 miles.

ASH FORK
<u>Rocky Creek Ranch</u>　　　　　　　　　　　Phone: 928-637-2727
Claudia Ladwig　　　　　　　　　　　　　　　or: 928-637-2289
PO Box 355 [86320] Directions: 1 mile from I-40, Call for Directions. **Facilities:** 6 indoor/outdoor pipe corrals, 3 outdoor stalls, and a 2 mare motel outdoor, feed/hay available, trailer parking with electric and water, 100'x300' arena, 60' round pen, overnight camping welcome, owner on site, open year-round, Call for reservations. **Rates:** $20 per day. **Accommodations:** Ash Fork Inn (1 mile).

BENSON
✶ <u>Circle R Ranch</u>　　　　　　　　　　　　Phone: 520-586-7377
Bobby Joe & Elly McFadden　　　　　　　Cell Phone: 520-249-2838
E-mail: mcfadden@theriver.com
2850 W. Drilling Road [85602] Directions: Call for details. 1 mile off I-10, Exit 299 - Easy off and on. **Facilities:** 12 acres pasture, 5 large pipe stalls with shelter and pasture; Full size arena available at extra charge. Dog kennels; 14 spaces for vehicles and horse trailers - $8.50 per night per vehicle (includes electric and water hookup). Local Vet and Farrier available. Tack and Feed storage. 10 minutes from Kartchner Caverns. 20 min. from historic Tombstone. 30 min. from Tucson, Sierra Vista, Bisbee and Willcox. 60 min. from Chiracuaha National Monument and the Mexican border. **Rates:** $8.50 per night per horse. U-feed and care. **Accommodations:** Private room PLUS 3 bedroom Bed & Breakfast available with large living room and full kitchen. $17.50 per person. Several motels within 3 miles.

ARIZONA

ALL OF OUR STABLES REQUIRE CURRENT NEG. COGGINS, CURRENT HEALTH PAPERS & OWNERSHIP PAPERS.

BENSON
J-Six Equestrian Center / Desert Breeze Arabians   Phone: 888-668-9088
Joyce & Jim Hostetter
3036 Williams Road [85602] **Directions:** 5 minutes off freeway, all paved roads. I-10 to J-Six/Mescal Exit South, 50 yds. turn left onto Williams Road, 8/10 mile on left side. **Facilities:** 15' x 30' pens with roof, 2 pastures 250' x 200', lighted 250' x 150' arena, feed/hay available, trailer parking. Kartchner Caverns 10 minutes from facility. **Rates:** $10 per day per horse. **Accommodations:** Holiday Inn, Best Western and others 5 to 10 minutes from facility, special rates available.

CASA GRANDE
Pinal County Fairgrounds   Phone: 520-723-5242
Terry Haifley
512 S. 11 Mile Corner Road [85222] **Directions:** Call for directions. Exit 194 off I-10, 7 miles to 11 Mile Corner Road, turn right, stable 1/4 mile on the right. **Facilities:** 80 covered 8' x 8' stalls, trailer parking, pasture/turnout available, no feed/hay. 122-acre fairgrounds, practice arena, rodeo arena, open all year. Call for reservations. **Rates:** $10 per day. **Accommodations:** Motels within 7 miles.

Rainfire Stables   Phone: 520-836-2602
Whendie (Speaks English), Antonio (Hablo Espanol)
E-mail: RainfireStables@msn.com
17500 West Hanna Road [85222] **Directions:** Call for directions. **Facilities:** 30 stalls. Large indoor barn with runs. Outside stalls 20'x20' and 24'x20'. 1 outdoor pen and 1 turnout 300'x300' with auto water, unlimited trail riding. Alfalfa/grass available. **Rates:** $15 per night, weekly rate available. **Accommodations:** 4-5 RV hook-ups (30 amp with water and sewer connections). Many hotels in Casa Grande 4 miles away.

DEWEY
Horse Breakers Unlimited   Phone: 928-632-5728
John & Marywade Gilbert
P.O. Box 687 [86327] **Directions:** I-17 to Rt. 169 exit. Call for directions. **Facilities:** Up to 75 - 18' x 28' covered stalls, large show barn, wash rack, 40' x 80' pens, feed/hay. **Rates:** $15 per day; $75 per week. **Accommodations:** Days Inn, Prescott Valley, 8.5 miles from stable.

DRAGOON
And the Horse you Rode in On   Phone: 520-826-5410
Debbie &Will Scott   E-mail: horseinn@vtc.net
2434 W. Dragoon Road [85609] **Directions:** Exit 318 on I-10, head South 7.3 miles, driveway on right **Facilities:** 9 21x21 and 1 21x42 stall, pellets/alfalfa/hay available, trailer parking, round pen. **Rates:** $15 per night. **Accommodations:** Morning Star B&B in Dragoon, Benson & Willcox motels-2 miles away.

ARIZONA

ALL OF OUR STABLES REQUIRE CURRENT NEG. COGGINS, CURRENT HEALTH PAPERS & OWNERSHIP PAPERS.

FLAGSTAFF
Hitchin' Post Stables,   Phone: 928-774-1719
Roger L. Hartman   928-774-7131
4848 Lake Mary Road [86001] **Directions:** Call for directions. **Facilities:** 10 - 12' x 12' enclosed stalls connected to a 12' x 24' run, 65'diameter covered round arena, grass/alfalfa mix hay, trailer parking. Vet and Farrier on call. No studs. **Rates:** $30 per day, $75 per week, $250 per month. **Accommodations:** Many motels in Flagstaff center, 4.5 miles west of stable.

MCS Stables   Phone: 928-774-5835
Oak Creek (89A) HC 30, Box 16 [86001] **Directions:** Exit 337 (Airport Exit) off I-17: 2.5 miles south on 89A. I-40 & Route 66 lead directly to I-17 & 89A. **Facilities:** 20 stalls, 12 - 16' x 20' outdoor pens, 9 horse barn, 8 pipe-enclosed pastures up to an acre in size, mix pellets & cubes. Horse boarding facility with manager on premises and vets and farriers on call. **Rates:** $25 per night. $285 for month. **Accommodations:** Motel 6, Ramada Inn & Fairfield Inn located 4 miles away in Flagstaff.

HEREFORD
Equi-Sands Training Center   Phone: 520-378-1540
Vicki Trout
9595 Kings Ranch Road [85615] **Directions:** Off State Hwy 92, call for directions. **Facilities:** 400 acres, 11 box stalls, 20 pens, 5/8 mile race track, 2-250' x 150' arenas, dressage arena, surrounded by 10,000 acres of state land. 2 miles from Coronado National Monument. **Rates:** $15 per day; $75 per week. **Accommodations:** Motels in Sierra Vista, 15 miles from stable.

HOLBROOK
Navajo County Fairgounds   Phone: 928-524-6407
Sam Pogue- Manager, Jose Villerreal-Groundskeeper
404 East Hopi [86025] **Directions:** Eastbound- Exit 285 east on Hopi to fairgrounds. Westbound-Exit 286 0r 287-South on Navajo Blvd, turn left on Hopi to fairgounds **Facilities:** 50 outdoor stalls available (12x12 Block, 12x12 chainlink, 14x14 metal), 250'x300' arena, trailer parking with water & electric for self-contained units at $5 per night **Rates:** $10 per night. **Accommodations:** Motels nearby.

PAGE 30 **ARIZONA**

ALL OF OUR STABLES REQUIRE CURRENT NEG. COGGINS,
CURRENT HEALTH PAPERS. & OWNERSHIP PAPERS.

KINGMAN
Mohave County Fairgrounds
Errol Pherigo
Phone: 928-753-2636
Fax: 928-753-8383
2600 Fairgrounds Blvd. [86401]
Directions: Call for directions.
Facilities: 230 - 10' x 10' outside covered box stalls. Open 24 hours. Water available year round. No feed/hay available. Open door policy. Night watch person. **Rates:** $10 per night.
Accommodations: Motels within 1 mile. RV space available $8.00 per day.

PAGE
Rope and Saddle Promotions, Inc.   Phone: 928-853-8718
Bea Thompson & Mike Bergner   928-645-2119
531 Haul Road, P.O. Box 3803 [86040] **Directions:** From Hwy 89 turn onto Haul Rd (Texaco station at intersection) Approximately one mile to entrance. Please call for other directions. **Facilities:** 50 stalls, outdoor pens 20"x40', round pens, arena, trail riding. Feed/hay available, trailer parking available. **Rates:** $12 per night $50 per week. **Accommodations:** Page is adjacent to Lake Powell with many motels and resorts.

PAULDEN
✱ Little Thumb Butte Bed & Breakfast   Phone: 928-636-4413
Ann Harrington   Fax: 928-636-4452
Web: www.littlethumb.net   E-mail: infoltb@commspeed.net
1252 Morgan Ranch Road (86334) **Directions:** Call for directions. **Facilities:** 10-12X12 covered block stall with runs and pipe fencing. Turn out 100'X 200'. Feed & Hay. Current health papers, negative coggins and up to date West Nile virus vaccination. **Rates:** $10 per night. **Accommodations:** B&B

PHOENIX
Royal Ranch   Phone: 625-879-8054
Bob & Margaret Shearburn, owners   or: 602-430-3171
504 W. Galvin St. [85086] **Directions:** From Interstate 17: Take the Carfree Hwy turnoff East to 7th Ave.(about 2 miles) Turn North on 7th Ave to Galvin st. then East to gate. **Facilities:** 40 indoor and outdoor stalls, feed/hay available, trailer parking, 24x24 turnout, indoor wash racks, indoor grooming area, fly & misting system, rubber mats, shavings, lighted arenas, bull pen, round pen, fenced and gated for security. **Rates:** $20 per day with feed, $15 without; $100 per week with feed, $85 without.

ARIZONA Page 31

ALL OF OUR STABLES REQUIRE CURRENT NEG. COGGINS, CURRENT HEALTH PAPERS, & OWNERSHIP PAPERS.

PRESCOTT
The Davis Ranch   Phone: 928-778-0895
Eileen Davis   or: 888-836-3211
1890 Pemberton Dr. [86305] **Directions:** Call for Directions **Facilities:** 3 12x12 indoor stalls, 12x12 outdoor stall, hay cubes available, trailer parking, Home of Davis Ranch combined driving team, driving lessons, mini stallion TeSono De Moro, Bed & Breakfast. **Rates:** $15 per day. **Accommodations:** Motels nearby.

SKULL VALLEY
✻ The Oasis Ranch   Phone: 928-442-9559
Bruce & Bonnie Jackson
P.O. Box 256 [86338] **Directions:** Located off U.S. Hwy 89. At Kirkland Junction take County Rd. 15 west to Kirkland. Turn right on County Rd. 10 (Iron Springs Rd.) Go approx. 7 miles to Skull Valley. Call for further directions. **Facilities:** 6 stalls, 90' x 120' arena turnout, alfalfa or grass 3x daily. **Rates:** $15 per day; $55 per week. **Accommodations:** Oasis Ranch is an inn with priority stabling given to guests. 2 self-contained B&B cottages. $125 per night. $550-$650 per week.

SONOITA
✻ Rainbow's End Bed & Breakfast and Gaited Horses   Phone: 520-455-0202
Charlie & Elen Kentnor   E-mail: ElenKentnor@compuserve.com
P.O. Box 717 [85637]   Web: www.gaitedmountainhorses.com
Directions: Take I-10 exit #281 (Sonoita, Patagonia Scenic Hwy 83) 24 miles south, crossroads Hwy 82, continue on Hwy 83, 7/10 mile, south side (right) look for our sign. **Facilities:** 25 stalls in two large barns. Spacious comfortable and safe. 17 stalls in large barn with 12' x 12' indoor with 100' runs and automatic waterers. Feed/hay available, Colorado grass for additional fee. Horse trailer parking only. Historic breeding ranch, (Rocky Mtn, Kentucky Mtn, Saddle horses). Trainer on premises to help with training needs or guide you on scenic tours for an additonal fee. **Rates:** $15 per night per horse. **Accommodations:** 4 bedroom / 4 bath B&B on premesis, stabling for B&B guest a priority, no RVs or camping permitted.

TUCSON
Cactus Country RV Resort   Phone: 520-574-3000
Nancy or Bob Iverson   800-777-8799
10195 S. Houghton Road [85747]   Web: www.arizonaguide.com/cactusco
Directions: I-10, exit 275 (Houghton Rd.), North 1/4 mile. East 1/2 mile to office. **Facilities:** Fenced pen will hold 12 "friendly" horses, no feed or hay. Short term stay only. Must have reservation during winter months. **Rates:** No cost for horse. **Accommodations:** Owner must be staying at RV Resort and have own RV or tent and pay for a full hook-up site.

ALL OF OUR STABLES REQUIRE CURRENT NEG. COGGINS, CURRENT HEALTH PAPERS, & OWNERSHIP PAPERS.

TUCSON
Catalina State Park   Phone: 520-628-5798
PO Box 36986, 11570 N. Oracle Road [85740] **Directions:** I-10, exit 240 (Tangerine Road), east 12 miles to Hwy 77, turn south and go 1 mile. Park is in foothills of Catalina Mtns. **Facilities:** 8 - 12' x 12' pipe corrals, no feed/hay but feed store within 5 miles. Must stay with horse, trailer parking available, showers; open 24 hrs/day, ranger security 24 hrs/day. Access to Coronado National Forest, known for birdwatching, wildlife, variety of trails for riding. **Rates:** $10 per vehicle. **Accommodations:** Motels within 6 miles.

✱ Rocking M Ranch, Bed & Breakfast   Phone: 520-744-2457 or 888-588-2457
Louis & Pam Mindes   Mobile: 520-444-0308   520-444-0306   520-906-3233
E-mail: lou@pamlou.com   Fax: 520-744-0824   Web: www.rockingmranch.net
6265 N. Camino Verde [85743] **Directions:** From Ina and I-10, west on Ina Road approximately 2 miles, south (left) on Camino Verde exactly 1 mile, road makes hard right west, continue a hundred feet and take first left. Follow to the end, can't miss the horse facility. Located just outside the Saguaro National Park in the Tucson Mountains. **Facilities:** 5 covered 13' x 12' 3 rail pipe corrals with auto water. Lighted arena, round pen and walker and miles of great riding. Reservations preferred, visits by appointment only, this is not a commercial facility. Boarding only for B&B guests. Veterinarians and farriers on call. **Rates:** Oct. 1 - April 30, $100 per night; May 1 - Sept. 30, $75 per night. Rates include breakfast. No charge for horse corrals. Guests responsible for animal feeding and cleanup. Cash or check no credit cards. **Accommodations:** B&B on premises.

WIKIEUP
✱ Bar 5 Cattle & Guest Ranch   Phone: 928-718-0000
Ron Robach; Pete & Molly Meyer   928-753-5285
10000 Chicken Springs Road [85360] **Directions:** Call for directions. **Facilities:** 6 50'x100' corrals w/turnout, 4 12'x14' pens, outdoor stalls, some covered. Alfalfa, hay and grain available, ample trailer parking. Trail riding, ranch work, cookouts. Dogs and cats not allowed unless strictly confined to a camper/motorhome. **Rates:** $15 per night, $75 per week. **Accommodations:** Trading Post Motel in Wikieup; 9 miles. RV hookups & two guest rooms at ranch.

WILCOX
Wilcox Livestock Auction   Phone: 520-384-2206
Sonny Shores, Scott McDaniel
Haskell & Patti Road [85643] **Directions:** Take the Wilcox Exit off of I-10. Call for further directions. **Facilities:** 600 outside pens, water & feed/hay available. Office open 8 A.M. to 5 P.M. On-call night manager's number posted on door. **Rates:** $10 per horse per night plus feed. **Accommodations:** Motels within 1 mile.

ALL OF OUR STABLES REQUIRE CURRENT NEG. COGGINS, CURRENT HEALTH PAPERS, & OWNERSHIP PAPERS.

WILCOX
<u>Dry Dock Horse Ranch</u>　　　　　　　　　　　Phone: 520-824-3359
Craig & Adel Lawson　　　　　　　　　　　　　　Fax: 520-824-3579
HCR 3 Box 5111 [85643] Directions: Call for directions. **Facilities:** 6 12'x12' indoor stalls, 6 12'x12' outdoor stalls, 1/4 acre turnout, feed/hay available, trailer parking. Located in the heart of Apacheria where Cochise, Geronimo, and Wyatt Earp rode. Guided trail rides available. Thousands of open acres to ride. **Rates:** $15 overnight, $10 per horse per night weekly. **Accommodations:** Guest Quarters available. RV hookup & dump. Bed & Breakfast 5 miles.

ARKANSAS

Missouri

Springdale
Blytheville
North Little Rock
Hackett
Conway
Ferndale
Hot Springs
Forrest City
Little Rock
Bismarck
Hope

Oklahoma
Texas
Louisiana
Mississippi
Tennessee

Towns Shown Are Stable Locations.
* onsite accommodations

ARKANSAS Page 35

ALL OF OUR STABLES REQUIRE CURRENT NEG. COGGINS, CURRENT HEALTH PAPERS & OWNERSHIP PAPERS.

BISMARCK (Hot Springs area)
✳ Bar Fifty Ranch, Phone: 888-829-9570
Bette & Julian Mckinney E-mail: admin@barfiftyranch.com
Web: www.barfiftyranch.com
18044 Hwy 84 [71929] **Directions:** Exit 91 off I-30, turn left, 9 miles on right. **Facilities:** 20 stalls with shavings, paddocks/turnout, feed/hay, trailer parking, hook-ups w/elec., sewer & water, nice trail riding. 20 min. to hot springs with mineral baths/fishing/boating/racetrack. 20,000 acres of trails. 50 site RV campground. Reservations required. **Rates:** $15 per night indoor, $10 covered, $5 outside pen. **Accommodations:** On premises, lodge with 2 bedrooms suites, sleeping 2-6; pool, hot tub & sauna, full country breakfast. $80 for double occupancy. MC/Visa.

BLYTHEVILLE
Circle S. Horse Motel Phone: 870-763-9203
Ronnie Self
3344 N. US Hwy 61 [72315] **Directions:** Exit 63 S off I-55 onto Hwy 61. Go 1 mile south. **Facilities:** 16 indoor box stalls, pasture area, camper hook-up. Stud facilities available. Tack shop. Feed/hay & trailer parking available. Call for reservations. Cash only. Check-out time 11 A.M. **Rates:** $20 per head/per night. **Accommodations:** Best Western 1 miles from stable; Comfort Inn & Days Inn in Blytheville, 4 miles from stable. No reservations accepted after 10PM.

CONWAY
Miss Toby's Horseback Riding Academy & Dance Ranch
Toby Hart, owner/operator Phone: 501-329-2233
Web: www.MissTobys.com E-mail: cjohnson@cyberback.com
255 E. German Lane [72032] **Directions:** North of Little Rock off of I-40. Call or visit website for further directions. **Facilities:** 10 stalls, 8 holding pens, 20 acres of pasture, grain/hay. lighted outdoor arena, round pen, walker, and wash rack. Unlimited trails in area. Children and pets welcome. Western pleasure lessons and rodeo lessons. Reservations appreciated. Cash or check. Owner lives on premises. **Rates:** $15 per day., $75 per week. **Accommodations:** Overnight loft accommodations on premises at the Dance Ranch, including shower, kitchen, internet access, phone, fax. Parking for trailers and big rigs. Water and electricity available. Numerous restaurant and motel lodging within 1 mile. Feed stores, Western stores, veterinarians, diesel, gas, and automotive repair shops. Farrier service available.

FERNDALE
The Barns Phone: 501-821-4422
William & Leslie Barns
2 Witness Tree Lane [72122] **Directions:** Located 13 miles from Little Rock. Please call for directions. **Facilities:** 3 indoor stalls, 4-acre pasture, large corral, feed/hay, camper parking if self-contained. **Rates:** $15 per night; $75 per week. **Accommodations:** Motels 13 miles away.

ARKANSAS

ALL OF OUR STABLES REQUIRE CURRENT NEG. COGGINS, CURRENT HEALTH PAPERS & OWNERSHIP PAPERS.

FORREST CITY
Roberts' Racing Stable — Phone: 870-633-9041
Stanley & Delia Roberts — Stanley Roberts' Cell: 870-261-2511
Roy Roberts' Cell: 870-261-4091
2685 Hwy 1 South [72335] **Directions:** Exit 241 on interstate 40 between Memphis TN and Littlerock AR. Take Hwy 1 south off I 40 , go 5 miles south. Barn 5 miles from I 40 on Hwy 1. **Facilities:** 74 indoor stalls, 12 x 12 indoor, 100 x 200 paddocks, 1-mile training track with starting gates, open runs & paddocks. Trainer of racehorses and breeder of thoroughbreds. Reservations required. **Rates:** $15 per day; ask for weekly rate. **Accommodations:** Motels within 3 miles of stable.

HACKETT
Daystar Arabians — Phone: 479-639-2401
Tom & Annetta Tinsman
26110 Hwy 45 South [72937] **Directions:** I-540, exit 14, south 16 miles, stable on right. **Facilities:** At least 6 - 10' x 10' and 10' x 12' indoor stalls, 40 acres of pasture, 60' x 100' indoor arena, feed/hay and trailer parking available. Mountain trail rides, complete training and horse education. Call for reservations. **Rates:** $10 per night; weekly rates on request. **Accommodations:** Motel and restaurant within 20 miles.

HOPE
Hope Fair Park — Phone: 870-777-7500
Paul Henley — Hours: 8-12 & 1-5, M-F
P.O. Box 596 [71802-0596] **Directions:** From I-30: Exit 30 for Hope, take right and another right at second traffic light. Go 2 blocks to Hwy 174 and take left. Go 5 blocks to Park Drive and take right. **Facilities:** 60 indoor stalls, paddock area, 100' x 180' outside covered arena, RV hook-ups with water & electric. Open 24 hours. Hope Feed Co. nearby. Rodeos & festivals held at Park. Temporary stalls set up during shows. **Rates:** $5 per night. Please put payment in drop box. **Accommodations:** Days Inn and Holiday Inn 2 miles from stable.

LITTLE ROCK
AM Ranch — Phone: 501-312-2818
Allen McKnight
13111 Colonel Glenn Road [72210] **Directions:** Exit 4 off I-430. Go west 1 mile on Colonel Glenn Rd. The facility is on the left. **Facilities:** 20 -12' x 12' indoor stalls; feed/hay and trailer parking are available. One acre pasture/turnout. **Rates:** $25 per night per horse. Negative coggins required. **Accommodations:** Room at facility is normally available or LaQuinta Inn on Shackleford Road 3 miles away.

ARKANSAS Page 37

*ALL OF OUR STABLES REQUIRE CURRENT NEG. COGGINS,
CURRENT HEALTH PAPERS & OWNERSHIP PAPERS.*

LITTLE ROCK
Fox Creek Farms   Phone: 501-225-9384
Kari Barber
4100 Bowman [72210]   **Directions:** I-430 to Exit 4. Go 1/2 mile west on Colonel Glenn to Bowman Rd. Turn right (north) on Bowman. Entrance approx. 1/2 mile on left. Drive to barn. **Facilities:** 23 indoor wood stalls, 8 indoor concrete block stalls, 3 outdoor stalls, many pasture/turnout areas, no feed/hay. Farm is closed and gates locked at 9 P.M. Must call and leave message if arrival after 9 P.M. No campers, tents, etc. allowed for sleeping on the grounds. **Rates:** $15 per night. **Accommodations:** Motel 6, LaQuinta, Holiday Inn 3-5 miles from farm.

N. LITTLE ROCK
McAdams Family Stables   Phone: 501-835-2205
Kenneth McAdams
10431 West Mine Road [72120] **Directions:** Call for directions. **Facilities:** 4 indoor stalls, arena, feed/hay available, trailer parking. Some notice if possible. **Rates:** $15 per night. **Accommodations:** Days Inn 4 miles from stable.

SPRINGDALE
Ryn-Char Horse Center   Phone: 479-750-3552
Karen Cook
4644 Hylton Road, WC 559 [72764] **Directions:** Call for directions. **Facilities:** 11 indoor stalls, 10 acres of pasture/turnout, indoor arena, outdoor lighted arena, roping facilities, feed/hay & trailer parking available. Vet. on premises & 2 farriers on call. **Rates:** $10 per night. **Accommodations:** Holiday Inn and Executive Inn 3 miles from stable.

Page 38 **CALIFORNIA**

Towns Shown Are Stable Locations.

* onsite accommodations

CALIFORNIA PAGE 39

ALL OF OUR STABLES REQUIRE CURRENT NEG. COGGINS, CURRENT HEALTH PAPERS & OWNERSHIP PAPERS.

ACTON
Broken Spoke Ranch   Phone: 661-269-3653
Greg Sachen & Jackie Tichenor
5525 Braeloch Street [93510] **Directions:** Call for directions. Easy access to Hwy 14. **Facilities:** 8 stalls, new 12' x 12' barn with stall mats & auto waterers, 12' x 24' run, 15' x 60' turnout, hay available, trailer parking available. **Rates:** $10 per day. **Accommodations:** Holiday Inn, Days Inn, E-Z 8 Motel, Palmdale Inn, Super 8 Motel all within 15-20 minutes.

ANDERSON
Tony Pochop Training Stables   Phone: 530-365-7759
Tony Pochop   530-378-2525
5729 N. Balls Ferry Road [96007] **Directions:** Take Deschutes Rd. exit off I-5 for 2.2 miles. Turn left on Balls Ferry Road. Go .4 miles. Stable is on right side of road. **Facilities:** 10 indoor stalls, 2 arenas, alfalfa hay, overnight parking for campers & motor homes available on grounds. Reservations appreciated but not required. **Rates:** $15 per night for stall, $10 for arena; weekly rates negotiable. **Accommodations:** Valley Inn & Knights Inn, Amerihost Inn only 2 miles from stable.

BAKERSFIELD
Galbraith's End of Road Ranch   Phone: 661-845-3013
Joanne and Dale Galbraith   Fax: 661-845-3043
E-mail: galbraithtrailers@hotmail.com
9010 Hermosa Rd (93307) **Directions:** 58 freeway to 184 South Weedpatch Hwy to Hermosa Rd. Turn east 1/2-mile North side of road. **Facilities:** 8- 12X48, 4- 20X20 covered pipe stalls. Feed/hay available. Possible trailer parking. We are AQAH breeder. Have stallions and horses for sale. Call ahead for reservations. **Rates:** $15 per night. **Accommodations:** Motel 8 on Breendage Ln, 3 miles off 58.

BARSTOW
Barstow Horse Motel & Ranch   Phone: 760-256-3671
Richard & Phyllis Dye   Web: www.horsemotel.homestead.com/HorseMotel.html
27702 Waterman St. [92311]   E-mail: horsemotelca@aol.com
Directions: Traveling south from Las Vegas area, 6 miles west of I-15 on Old Hwy-58 or 3 miles from down town Barstow City on Old Hwy 58. From New Hwy 58, take Lenwood Road north to Old Hwy 58 and go east 4 miles. We are on the east side of Old Hwy 58. **Facilities:** 13- 20' x 20' stalls plus 24' x 24' stud pen, 11 outdoor pens, 100' x 120' turnout available, work area. Alfalfa hay ($2.00 per feeding), water. Electric hook-up for campers and motor homes at additional charge of $20 per night. Dogs need to be kept on leash or tied up. **Rates:** $10 per night; $35 per week; $150 per month. **Accommodations:** Many motels 3 miles form town.

CALIFORNIA

ALL OF OUR STABLES REQUIRE CURRENT NEG. COGGINS, CURRENT HEALTH PAPERS & OWNERSHIP PAPERS.

CARMEL VALLEY
JM Farm   Phone: 831-659-2553
Jon & Mary Sutherland
550 W. Carmel Valley Road [93924] **Directions:** From I-10: Beaumont Ave. exit, go north 3 miles; turn left on Vineland at stop sign. Stable is 1/2 mile on right. **Facilities:** 6 indoor stalls, 9 holding pens, pasture, camper hook-up with water & electric, veterinarian available. Morgan breeder, trainer & sales. Transports horses nationwide. **Rates:** $15 per night. **Accommodations:** Motels within 3 miles.

CHERRY VALLEY
Gee Jay Ranch   Phone: 951-845-5859
Gerry & Judi Brey   Fax: 951-769-9839   Cell: 951-318-3449
38660 Vineland [92223] **Directions:** From I-10: Beaumont Ave. exit, go north 3 miles; turn left on Vineland at stop sign. Stable is 1/2 mile on right. **Facilities:** 9 indoor stalls, 13 holding pens, pasture, camper hook-up with water & electric, veterinarian available. Morgan breeder, trainer & sales. Transports horses nationwide. **Rates:** $15 per night. **Accommodations:** Best Western and Budget Inn in Beaumont, within 3 miles.

GOLETA
Horseman's Hangout   Phone: 805-685-4440
Marcia Nelson   Fax: 805-685-9020
10920 B2 Calle Road [93117]   E-mail: mnhorsin@aol.com
Directions: Highway 101 take El Capitan Ranch Road exit, 10 miles from Santabarbara, 30 miles from Buellton. Reservations please and call if plans change. **Facilities:** 12 - 12' x 12' box stalls. Some outdoor pens with shade. Full size arena and outdoor round pens with shade. Natural Horsemanship/ Balanced Centered Riding. Clinics & activities, trails and beach access. Nearby summer camps. Overnight rates $10 to $25 depending on service. **Accommodations:** Full service camp ground next door (elcapitancanyon.com) and lots of motels 10 to 30 miles away.

IONE
Sunnybrook Ranch   Phone: 209-274-2680
Jim & Lori Mote   Web: www.sunnybrookequineappraisals.com
Fax: 209-274-0392   E-mail: sunnybrook@cdepot.net
9401 Brook Ranch Road W. [95640] **Directions:** Easy access off of Hwy 88. Call for directions. **Facilities:** 10 indoor stalls, 5 outdoor paddocks, 2-4 acres pasture/turnout, wash rack, riding arenas, covered arena, parking for large rigs. Tack repair and used tack for sale. Professional horse appraiser, young horses for sale. Call for availability. **Rates:** $15 per night; $95 per week. **Accommodations:** 5 miles from Gold Country towns of Ione, Sutter Creek, and Jackson. Many motels to choose from.

CALIFORNIA Page 41

ALL OF OUR STABLES REQUIRE CURRENT NEG. COGGINS, CURRENT HEALTH PAPERS & OWNERSHIP PAPERS.

JULIAN
Stagecoach Trails RV, Equestrian & Wildlife Resort
Phone: 760-765-2197
Robert & Michelle Crofts   Web: www.stagecoachtrails.com
7878 Great So. Overland Stage Route of 1849 [92036] Directions: Call for best directions. **Facilities:** 70 24x24 outdoor stalls, hay available, trailer parking, 85' round pen, RV and horse camp resort, pool, store, lodge for groups, unlimited riding for all levels, rental trailers. NO Stallions. **Rates:** $5 per night. **Accommodations:** We have RV spots, tent spots and rental trailers.

LOS OSOS
Sea Horse Ranch, BBG   Phone: 805-528-0222
Barbi Breen-Gurley
2566 Sea Horse Lane [93402] Directions: From San Luis Obispo off Hwy 101: Take los Osos Valley Rd. Go 12 miles to left onto Doris Drive, take Doris to Highland Drive; Right on Highland, Follow dirt road up to our sign "Sea Horse Ranch" **Facilities:** 4 12'x24' Pipe corrals with shelter, Feed/Alfalfa available, very limited trailer parking, we are primarily a dressage training stable; we also have access to the Montana de Oro State Park and the ocean beaches. **Rates:** $20 per night. **Accommodations:** Lots to choose from in in nearby Morroray (10 miles) or San Luis Obispo (12 miles).

LOST HILLS
Lost Hills KOA   Phone: 661-797-2719
Scott Neufeld
14831 Warren Street, I-5 and Hwy 46 [93249] Directions: West on Hwy 46 from I-5. Go 1/8 mile, south at Carl's Jr. KOA is 100 yds. **Facilities:** 4 outdoor stalls with trees, each holds 4 horses: 3 - 12' x 24' and 1 - 12' x 24' for stallions, trailer parking, no feed/hay. Must clean up area before leaving. Easy walk to restaurants & service stations. By reservation only: call 1-800-562-2793. **Rates:** $10.50 per night; $50 per week. **Accommodations:** Motel 6 and Days Inn, both 200 yds from stable.

MILL VALLEY
Miwok Stables   Phone: 415-383-6953
                                 415-381-0529
Gabino Saldona, mgr.
701 Tennessee Valley Road [94941] Directions: Take Hwy 101 to exit for Hwy 1 Stinson Beach; take Hwy 1 for 1/4 mile, past Holiday Inn; turn left on Tennessee Valley Road, go 2 miles to end. **Facilities:** 42 outdoor pens or pasture, alfalfa/oats, trailer parking. Public riding facility with excellent trails. Located in Golden Gate National Recreation Area, 680 acres of trails; 10 minutes from San Francisco in Marin County. **Rates:** $15 per night. **Accommodations:** Holiday Inn in Mill Valley 2.5 miles away.

Page 42 CALIFORNIA

ALL OF OUR STABLES REQUIRE CURRENT NEG. COGGINS, CURRENT HEALTH PAPERS & OWNERSHIP PAPERS.

MILPITAS
Diamond W Ranch   Phone: 408-262-4163
Ernie Wool
Weller Road [95035] **Directions:** Take Calaveras Road off of 680. Call for further directions. **Facilities:** 100 indoor stalls, 80' x 220' indoor arena, pasture/turnout available. English & Western riding lessons, cutting & reining lessons. Call for availability. **Rates:** $15 per night, feed included; $185 per month. **Accommodations:** Motels located 2 miles from ranch.

QUINCY
✱ **New England Ranch**   Phone: 916-283-2223
Barbara Scott, Rick Tegeler
2571 Quincy Junction Road [95971] **Directions:** 3 miles from Hwy 70 along Chander Road to corner of Quincy Junction Road. **Facilities:** 12' x 12' indoor stalls, 20' x 20' outdoor stalls, 15+ acres of pasture, 50' x 100' turnout, feed/hay available, trailer parking. Riding arena, full-care boarding facility, access to Plumas National Forest riding trails. **Rates:** $15 per night. **Accommodations:** Bed & breakfast (2 rooms) and bunkhouse trailer (sleeps 5) on premises, $60-$95 per night includes gourmet breakfast.

RED BLUFF
Bar M Ranch   Phone: 530-527-2107
20105 Callahan Rd   E-mail: barmrch@tco.net
Directions: 5 miles off I-5. Call for directions. **Facilities:** 6 large indoor stalls. Feed/hay and trailer parking, big rig parking, pull through access for large rigs, hookup available, arena, round pen. **Rates:** $10-20 per night. **Accommodations:** Many hotels and restaurants 3 miles away.

REDDING
3 D Ranch   Phone: 530-549-3049
Vicki Donovan
20567 Conestoga Trail [96003] **Directions:** Call for directions. **Facilities:** 6 12x12/12x24 stalls, feed/hay available, trailer parking, 60' round pen. Located just minutes from I-5, 3D Ranch is a full service horse facility. We raise Icelandic horses, train all breeds, board long term and welcome travelers with horse. RV hookups at no extra charge **Rates:** $20 per night; $100 per week. **Accommodations:** Motel 6, Ramada, Holiday Inn Express- 3 miles

SACRAMENTO
Cracker Jack Ranch   Phone: 916-363-4309
Lew & Jeanee Conner   Voice Mail: 916-441-8179
10004 Jackson Road [95827] **Directions:** Please call for directions: **Facilities:** 8 stalls in barn, 4 stalls with outside corrals, paddocks, auto waterers, arena, turnout pens. Parking for campers if self-contained. Stallion service for thoroughbreds & quarter horses. Horse training available. Call for reservations. **Rates:** $15 per night. **Accommodations:** Howard Johnson's located 4 miles from stable.

CALIFORNIA Page 43

ALL OF OUR STABLES REQUIRE CURRENT NEG. COGGINS, CURRENT HEALTH PAPERS & OWNERSHIP PAPERS.

SAN DIEGO
Clews Horse Ranch Phone: 858-755-5022
Christian Clew
11911 Carmel Creek Road [92130] **Directions:** From I-5 take Carmel Valley Road Exit. **Facilities:** 80-horse ranch with ocean views. 24' x 24' pipe corrals, auto waterers, auto feeders, lighted arena, bull pens, wash racks, miles of trails. Breeder of old-style quarter horses. Stallion is "Chubby Lobos Last." Call ahead for availability. **Rates:** $15 per night. **Accommodations:** Motels within 1 mile of stable.

SANTA BARBARA
✶ **Rancho Oso Guest Ranch & Riding Stables**
Bill Krzyston, Manager Phone: 805-683-5686
Lil Rosen, Reservations Fax: 805-683-5111
3750 Paradise Road [93105] Web: www.rancho-oso.com
Directions & Rates: See web site (www.rancho-oso.com) or call us. **Facilities:** 24 outdoor pipe corrals, 8 indoor box stalls, arena, round pens, hay twice a day, trailer parking. Access to Los Padres National Forest trails. Open year round. Complete camping facilities plus overnight accommodations in cozy cabins or covered wagons. Weekend meals, hot showers, pool & spa. Guided trail riding on their horses.

SANTA MARGARITA
✶ **Santa Margarita K.O.A.** Phone: 805-438-5618
Rex Jacobson, owner
4765 Santa Margarita Lake Road [93453] **Directions:** Santa Margarita exit from Hwy 101. Hwy 58 to Pozo Rd., follow signs to lake about 8 miles. **Facilities:** 4 outdoor pole corral stalls on 68-acre campground. Riding trails & beach riding nearby. Great fishing. Tent sites, full hookup sites, Kamping Kabins, pool, country store, & laundry. **Rates:** $7 stall (2 horses) per night, 7th night free. **Accommodations:** On premises.

Page 44 **COLORADO**

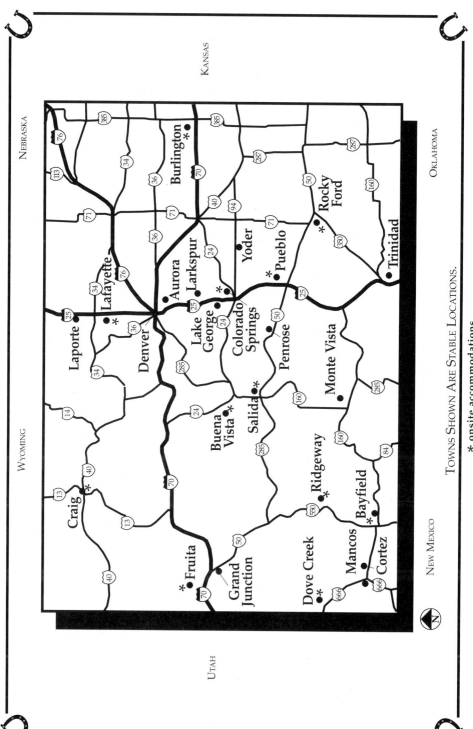

COLORADO Page 45

ALL OF OUR STABLES REQUIRE CURRENT NEG. COGGINS, CURRENT HEALTH PAPERS & OWNERSHIP PAPERS.

AURORA
Kenlyn Stables   Phone: 303-807-0062
Linda Fisher   Fax: 303-366-7454
E-mail: Kenlyn@idcomm.com   Web: www.kenlynarabians.com
1000 Salida Street [80011] **Directions:** I-70 Airport Blvd South exit #285. 2 miles south to 10th Ave. East 3 blocks to Salida. We're on the corner. I-225 6th Ave east 2 miles to Airport Blvd. Left 1/4 mile to 10th Ave. Right 3 blocks to Salida. **Facilities:** 16 indoor and outdoor stalls, four 36' x 60' turnout areas, trails, indoor and outdoor arenas. Grass and alfalfa available. This is a show barn - clean and safe. Also Arabian breeding farm specializing in endurance riding, pleasure, and show. Endurance cross country horses along with halter & show winners. Studs available and horses for sale. Lessons & sales for all ages. Call in advance. **Rates:** Varies from indoor to outdoor stalls. **Accommodations:** Holiday Inn on I-70 and Travelers Inn 2 miles from stable in Aurora.

BAYFIELD
✱ **Horseman's Lodge**   Phone: 970-884-9733
Andras & Lorraine Westwood, Owners   or: 800-715-6343
7100 County Road 501 [81122] **Directions:** US 160 to Bayfield. 7 miles on Rt. 501 on right side. **Facilities:** New arena and stalls coming in 2003! Run in shed and corral, Lodge has 9 rooms, 7 with kitchenettes with daily, weekly and monthly rental. Pets welcome. **Rates:** $10 per night per horse. **Accommodations:** 4 new deluxe cabins on site.

BUENA VISTA
✱ **Wapiti Run Trakehners**   Phone: 719-395-8543
Judy & Jim Moore   E-mail: dialogue@amigo.net
17900 Vista Drive [81211] **Directions:** From Hwy 24: Call for directions. **Facilities:** 1/2- to 2-acre paddocks with run-in log barns & automatic heated waterers, 3 paddocks, 2 very large pipe pens, 2 outdoor stalls, grass hay available. This is a full training facility with indoor & outdoor arenas & a cross-country course and lessons and training in dressage offered. Located in the Upper Arkansas Valley at the base of 14,000 ft. Mt. Columbia with direct access to the Colorado trail. Pets welcome if controlled. **Rates:** $18 per calendar day without feed. **Accommodations:** 2 furnished studio apartments on premises available for rent.

COLORADO

ALL OF OUR STABLES REQUIRE CURRENT NEG. COGGINS, CURRENT HEALTH PAPERS & OWNERSHIP PAPERS.

BURLINGTON
✷ **JB Horse Motel, B&B**　　　　　　　　　　Phone: 719-346-8217
Joan Chandler, Owner
19800 Hwy 385 [80807]. **Directions:** From I-70: Exit 438- Go west on Rose Ave. 1 1/2 blocks; Turn right on 8th St (AKA 385N) to RR tracks (approx 6 blocks) and continue north. Stable is 1.3 miles past railroad tracks on the right. **Facilities:** 7 indoor stalls, 6 stalls with pipe runs, 150' x 150' & 80' x 80' pastures, numerous corrals & pens, outdoor arena with lights. Reservations preferred. **Rates:** $20 per night. Electric and water hookups $5.00. **Accommodations:** Super 8 and Chaparral Budget Host about 2 mile away.

COLORADO SPRINGS
✷ **H2 Stables**　　　　　　　　　　　　　　Phone: 719-495-2338
Ed & Arlene Housley, Owners　　　　　　Web: www.H2Stables.com
E-mail: H2Stables@mindspring.com
6665 Walker Road [80908] **Directions:** I-25, exit 161, Hwy 105 E. To Hwy 83, the 3 miles on Walker Road. **Facilities:** Summer-9 12x12 indoor, 3 15x25 outdoor stalls. Winter-4 12x12 indoor, 3 15x25 outdoor stalls, hay/grass available, trailer parking, 2 turnouts, trail riding within minutes at Air Force Academy, Fox Run Regional Park, Black Forest Fark. **Rates:** $18 per night. **Accommodations:** Bed & Breakfast on premises (2 bedrooms with bath). reservations necessary. No Smoking.

CORTEZ
Jones Livery　　　　　　　　　　　　　Phone: 970-565-9639
Jerry & Carol Jones　　　　　　　　　　　Web: www.joneslivery.com
E-mail: jcjoneslivery@aol.com
20225 Country Road G, P.O. Box 1378 [81321] **Directions:** Located south of Cortez. See state directional sign on 491/160 South. County Road G is across from the DOT Inspections Station/truck stop. Proceed west to the traffic light on County Road G, 2.5 miles past Cortez Air Port and turn right at sign. Owners will meet travelers at the county road during night hours. **Facilities:** Overnight, short stays, and long-term stabling available. 12'x12' matted indoor stalls, each with a 12'x 20' outdoor corral. Includes professional-sized open arena, large covered lighted arena, and exercise pen. Trailer parking area can accommodate big-rigs and large trailers. Arenas are available for competition, clinics, and special events. Facility is private and secure. Owners live on property. **Accommodations:** Major Motels 6 miles from stable.

✷ CRAIG
Two Shoes Horse Ranch　　　　　　　Phone: 970-824-0105
Bill & Dona Shue　　　　　　　　　　　　Cell: 970-629-5760
E-mail: twoshueshr@earthlink.net　　　Web: www.twoshueshr.8k.com
696 County Road 22 [81625] **Directions:** South of I 80. 90 miles on Highway 13 North of I 70 on Highway 13, 95 miles. 6 miles N. on Highway 13 from US 40. **Facility:** 4 covered outdoor 10'x12' stalls. Turnout area and pasture 40'x 50'. Feed and hay available. Trailer parking available. Can accommodate big rigs and large trailers. **Rates:** $15 per night. Call for reservations. **Accommodations:** B&B on site. Motels in Craig 5 miles from stables.

COLORADO PAGE 47

ALL OF OUR STABLES REQUIRE CURRENT NEG. COGGINS, CURRENT HEALTH PAPERS & OWNERSHIP PAPERS.

DOVE CREEK
✻ **Sun Canyon Ranch** Phone: 970-677-3377
Mac McMahon Toll Free: 866-737-3377
Web: www.suncanyonranch.com   E-mail: mail@suncanyonranch.com
02082 Rd 8 [81324] **Directions:** 6 1/2 miles North of Dove Creek on Guyrene/ Rd 8 off of Hwy 491/666. **Facilities:** 7 indoor, 10 outdoor stalls, outdoor paddocks. Round pen, outdoor arena, heated wash stalls 12 RV sites with 50 amps, B&B room in stable, lodge sleeps 20+, great views, Dolores River Canyon trailhead great riding, massage, hot tub, guided rides, lessons. Many area attractions. **Rates:** $22.50/18.00 per night. 20% discount on weekly rate. **Accommodations:** We offer lodging.

FRUITA
✻ **Valley View Bed & Breakfast** Phone: 970-858-9503
Lou A. Purin
888 21 Road [81521] **Directions:** Exit 26 off I-70. One mile west to 21 Road, go north on 21 Road one mile. Stable on right side of road. **Facilities:** 9 outdoor stalls, 50' round pen, trailer parking. Please call in advance. **Rates:** $10 per night. **Accommodations:** Valley View B & B on premises; Westgate Motel, Grand Junction, 2 miles at Exit 26.

FORT COLLINS
Kenlyn Stables North Phone: 303-566-1986
Kate Fisher Fax: 970-484-8714
Web: kenlynarabians.com   E-mail: katfish242@yahoo.com
3317 Michaud Lane [80521] **Directions:** Call for directions from I-25. From Hwy 287 in LaPorte, south on Overland Trail 1 mile to corner of Michaud Ln. **Facilities:** 10 12'x150' stalls with shelter. Feed/grass available. Trailer parking available. 80'x 80' turnout, 100 acres of pasture full time boarding with access to open space trails. **Rates:** $15 per night, $90 weekly, $250 monthly. **Accommodations:** Many motels within 5 miles of stable.

GRAND JUNCTION
Alamar Stable Phone: 970-523-1445
Kim & Ben Shipard
3363 1/2- C Road [81526] **Directions:** 6 miles from I-70, Call for directions. **Facilities:** 6 stalls (12x12 indoor stalls with 12x24 runs, 20x40 outdoor stalls), alfalfa available, trailer parking, 2 round pens, 100x100 turnout. Full service training, boarding and breeding facility. **Rates:** $15-$20 per night. **Accommodations:** Best Western (5 miles).

LAFAYETTE
✻ **Mary Bradley** Phone: 303-828-4372
1375 N. 111th St. [80026] **Directions:** 10 minutes from Boulder. Call for directions. **Facilities:** 10 indoor/outdoor stalls, holding pens, pasture, arena, and riding trails. **Rates:** $12 per night. **Accommodations:** B & B on premises.

COLORADO

ALL OF OUR STABLES REQUIRE CURRENT NEG. COGGINS, CURRENT HEALTH PAPERS, & OWNERSHIP PAPERS.

LAKE GEORGE
✳ <u>Mule Creek Outfitters/M Lazy C Ranch</u>　　Phone: 800-289-4868
Randy and Brenda Myers　　Fax: 719-748-3250
Web: mlazyc.com　　E-mail: mlazyc@aol.com
Box 461 [80827] **Directions:** Hwy 24 west 5 miles past Lake George (between mile marker 260&261). Watch for yellow sign on the North side of road (35 miles west of Colorado Springs). **Facilities:** 15-16X16 indoor/outdoor stalls. Trailer parking available. The ranch is surrounded by 25,000 acres of Nat'l Forest. Great vacation spot. Health and coggins required. **Rates:** $10.00 per night per horse. **Accommodations:** Cabins available at the ranch.

LAPORTE
<u>Copper Top Acres</u>　　Phone: 970-221-4382
Lee & LoraLee Carter
4625 Kiva Dr. [80535] **Directions:** Located 6 miles NW of Ft. Collins, CO on Hwy 287. At intersection of County Road 54 G and Hwy 287 turn east on 54G off Hwy/287. Take first left on Pueblo, then left on Kiva Drive, Second home on the right. **Facilities:** New 3 stall Morton Barn with 12'x42'runs and 120'x90' working arena, wash rack and hay available, trailer hookup with water and electricity. Overnight accommodations at owner's home. Air conditioned room with double bed and trundle twin beds and private bath. located in close proximity to Rist and Poudre Canyons and Lory State Park **Rates:** $20 per day, $100 per week. **Accommodations:** Holiday Inn, Best Western Kiva Inn-Ft. Collins (6 miles). KOA campground and Heron Lake RV Park (2 to 3 miles).

LARKSPUR
<u>Rocky Mountain Training Center</u>　　Phone: 303-681-3237
Bobbi Richine　　Fax: 303-681-3238
E-mail: bobbi@rmtc.net　　Web: RMTC.net
6203 Valley High Rd [80118] **Directions:** Coming from North take exit 173 off of I-25. From South exit 172. Go into town of Larkspur and turn right at stop sign. Go 3 miles to the yield, turn right. Go 1.1 miles to Perry Park Ranch, turn left on Red Rocks Rd. Go 2 miles road splits, stay left for another 1.2 miles pavement ends, and keep going 1 mile sign on gate. **Facilities:** 30 outside with shelter, 16 indoor, 3 round pens and outdoor arena, pasture and turnout. Feed/hay and trailer parking available. Lessons and trail riding, ranch boarders national forest. **Rates:** $15 per night, $75 per week.**Accommodations:** Motels 17 miles away.

COLORADO Page 49

ALL OF OUR STABLES REQUIRE CURRENT NEG. COGGINS, CURRENT HEALTH PAPERS, & OWNERSHIP PAPERS.

LARKSPUR
✱ Spring Canyon Ranch, Camping & B&B Phone: 303-681-3237
Web: www.rmtc.net Barn Phone: 303-681-2942
E-mail: bobbi@rmtc.net Cell Phone: 303-808-9730
6203 Valley High Rd. [80118] **Directions:** (35 miles south of Denver) Exit I25 at #172 or #173. Go west into town of Larkspur. Turn right at stop sign. Proceed 3 miles to Hwy 105 and turn right. Go 1.3 miles to Perry Park Ranch subdivision. Turn left on Red Rocks Road. Go 2 miles to split in road. Stay left. Go 1.2 miles. Pavement ends. Stay with gravel road bearing left when in doubt. Road takes you right to our gate. **Facilities:** Pens, pasture, indoor stalls and covered arena. **Rates:** Horses $15 per day. **Accommodations:** A beautiful mountain valley with hundreds of miles of America's most scenic riding trails. Sleep in authentic Sioux tipi, your camper or our beautiful ranch house bedrooms. Camper hookups & shower available.

LONGMONT
CR Livestock & Animal Care, Inc. Phone: 303-651-7193
Rick & Chris Foster
757 Weld County Road 18 [80504] **Directions:** I-25 West, 3 miles on Hwy 52 to County Line Rd. Turn north - 2 miles to W.C.R. 18. **Facilities:** 3 to 4 enclosed stalls in heated barn, round pen, small indoor arena, outdoor arena, plus 1 port-a-stall barn w/access to pasture. Horse transportation - Colorado P.U.C. 24-hour reservation required. **Rates:** $15 per night. **Accommodations:** Budget Host & Super 8 in Del Camino, 8 miles away.

MANCOS
Samora's Horse Shoeing & Boarding Phone: 970-533-7500
Johnnie Samora
8730 CR 39 [81328] **Directions:** 1/4 mile off Hwy 160. 30 miles west of Durango on left or 17 miles east of Cortez. Turn right on CR 39. (Call for more info). **Facilities:** Stalls, arena & hot walker. **Rates:** $25 Per night per horse. Six miles from Mesa Verde National Park.

MONTE VISTA
Greenie Mountain Stable Phone: 719-852-5269
✱ Julie Burt
5041 Hwy 15 South [81144] **Directions:** From Jct. 160 & 285 in Monte Vista turn south onto Hwy 15. Follow Hwy 15 for 5 miles. Driveway is on west side, just past the 5 south road. **Facilities:** 13 12'x12' stalls, 7 w/pipe runs; runs vary in size. Grass/hay included in price, turnout available, trailer parking. Indoor and outdoor arenas. Close access to trails. **Rates:** $10 per night. **Accommodations:** Lodging available on grounds, call for availability. Other lodging 5 miles away in Monte Vista.

PAGE 50   **COLORADO**

ALL OF OUR STABLES REQUIRE CURRENT NEG. COGGINS, CURRENT HEALTH PAPERS, & OWNERSHIP PAPERS.

PENROSE
Caballo Casa   Phone: 719-372-6182
Jim & Nancy McEnulty
60921 E. Hwy 50 [81240] **Directions:** 28 miles west of Pueblo & 8 miles east of Canon City. Call for directions. **Facilities:** 4 box stalls, large round pen, trailer parking, hay only. **Rates:** $15 per night; $75 per week. **Accommodations:** Motels in Canon City, 8 miles from stable.

PUEBLO
Colorado State Fairgrounds   Phone: 800-876-4567
Ask for Horse Show   After 5, call 719-561-8484 or 719-561-8489
[81004] **Directions:** Central Avenue exit (97A) off I-25. Call for directions. **Facilities:** 400 stalls (not available if horse show in progress), feed/hay available before 5 p.m., trailer parking near stalls, camper/RV hook-ups. Call for reservations. **Rates:** $10 per night. **Accommodations:** Motels within 10-15 minutes of fairgrounds.

* Five Star Ranch & Equestrian Center, LLC   Phone: 719-382-5601
Doug Proctor, Manager   Web: www.fivestarranch.com
18550 Midway Ranch Road [81008] **Directions:** From the North; take I-25 to exit 119, turn left under freeway, right on Frontage Rd. Go North 1/2 mile. From the North; Take I-25 to exit 122, turn right, then left on to Frontage Rd. Follow South. Turning right after dip in road. **Facilities:** 130- 12x12 and 12x20 foaling stalls. Feed/hay, trailer parking available. 1-5 acre pastures, 20x40 turnouts. Full breeding program, lesson program, english & western show clinics and special events. **Rates:** $15-20 per night. **Accommodations:** On-Site guest rooms, RV hookup(electric).

Fountain Valley Stable   Phone: 719-545-8350
Lena Fox
2580 Overton Road [81008] **Directions:** Call for directions. **Facilities:** 9 indoor box stalls, 20 outside pens, 2 arenas, breaking pens, no feed/hay, trailer parking. **Rates:** $10 per night; $75 per week. **Accommodations:** 5 miles away.

RIDGEWAY
* Angel Ridge Ranch   Phone: 970-626-4395
Denise Fisher   E-mail: angelridge@frontier.net
177 County Road #10 [81432] **Directions:** Call for Directions. 25 miles South of Mentrose, CO off of Hwy 550. **Facilities:** 14 10'x12' stalls, outdoor paddocks with shelters. feed/grass & alfalfa available, trailer parking, 52 acres divided into 3 separate pastures with water access and shelter. Instruction offered, spectacular trails, vet services close by, year-round activities galore! Best destination vacation you will ever experience. **Rates:** $20 per night. **Accommodations:** B & B onsite premesis.

COLORADO Page 51

ALL OF OUR STABLES REQUIRE CURRENT NEG. COGGINS, CURRENT HEALTH PAPERS, & OWNERSHIP PAPERS.

ROCKY FORD
✱ **McComber Working Ranch** Phone: 719-980-2987
Jack & Sylvia McComber, Owners
21777 Hwy 71 [81067] **Directions:** Call for Directions. **Facilities:** 4 12X16 outdoor stalls plus 2 large corrals, feed/hay available at extra cost, trailer parking, turnout arena, 25 years experience in horse training, showing, breeding and cattle ranching. Facility is functional not fancy. Close to Comanche National Grasslands for trailriding. Clean your own stalls, Negative Coggins and Health papers, If traveling with dogs or cats, our rules apply. **Rates:** $10 per night. **Accommodations:** Charming cabin onsite, or motels in nearby LaJunta.

SALIDA
✱ **Beddin' Down Bed, Breakfast & Horse Hotel** Phone: 719-539-1815
Carolyn Sparkman   Web: www.BeddinDown.com
E-mail: Carolyn@BeddinDown.com
10401 CR 160 [81201] **Directions:** Hwy 50 thru Salida to Poncha Springs. Turn right. Go less than 1/4 mile to Hwy 285 North. Turn right. Go 2 miles to CR 160. Turn right. We are 1/2 mile down on the righ. **Facilities:** 16, 12' x 24' or 8, 12' x 48' stalls. Large riding arena and turnout. Cutting horse clinic (twice a year). Trail clinics (twice a year). Coggins, health and ownership papers for out of state. **Rates:** $15/20 per night. **Accommodations:** Beddin' Down is a B&B as well as a Horse Hotel.

✱ **The Tudor Rose** Phone: 800-379-0889 or 719-539-2002
Jon & Terre' Terrell   Web: www.thetudorrose.com
6720 County Road 104 [81201] **Directions:** Turn south onto County Rd 104 (Paradise Road) off of Hwy 50 on east side of Salida. Stable is at end of Hwy 104. **Facilities:** 4 - 10' x 10' indoor box stalls, 3 - 100' x 400' paddocks, feed/hay, trailer parking. No smoking in house or barn. Access to Rainbow Trail and Colorado Trail and BLM land for riding, guided trips available. White water rafting, mountain biking, hiking, cross country and Nordic skiing, fishing, hunting, golf available in area. **Rates:** $9.50 per night, includes hay. **Accommodations:** Beautiful Bed & Breakfast on premises.

TRINIDAD
LP Boarding Phone: 719-846-7094
Denise Pfalmer   E-Mail: d_pfalmer@hotmail.com
101 W. Indiana [81082] **Directions:** Take Exit 15 off of I-25. Turn west on Goddard & proceed to first stop sign. Turn right on Arizona & go to second stop sign. LP Boarding is across from stop sign. **Facilities:** 2 indoor box stalls, 6-8 outdoor stalls w/ 20' x 30' paddocks, 100' x 100' & 50' x 100' turnouts. Will also accept sheep/cattle and other livestock. Feed/hay available. Please call for reservations. **Rates:** $20 per pen. **Accommodations:** Hook-ups available. Hotels nearby.

COLORADO

ALL OF OUR STABLES REQUIRE CURRENT NEG. COGGINS, CURRENT HEALTH PAPERS, & OWNERSHIP PAPERS.

YODER
Silver Creek Ranch   Phone: 719-478-5007
Rosalee or John Hopkins
3755 N. Ramah Hwy. [80064] **Directions:** Call for directions. 3 miles west and 3.25 miles north of Rush off Hwy. 94. **Facilities:** 12 stalls, 300-acre pasture, wash rack, heated barn, outside arena, round pen, special stud pen, several lean-tos and runways with tall fencing. Call for reservations. **Rates:** $15 per night; weekly and monthly boarding available. **Accommodations:** Motels in Calhan (20 miles away).

NOTES AND REMINDERS

CONNECTICUT

Towns Shown Are Stable Locations.

* onsite accommodations

CONNECTICUT   PAGE 55

ALL OF OUR STABLES REQUIRE CURRENT NEG. COGGINS,
CURRENT HEALTH PAPERS & OWNERSHIP PAPERS.

BETHANY
Bittersweet Farm   Phone: 203-393-2586
Lance Wetmore
325 Amity Road [06524] **Directions:** 7 miles from I-84. Call for directions. **Facilities:** 35 indoor stalls, 72' x 120' indoor arena, 60' x 88' outside arena, 2/10 mile track, outside paddocks, 2 large fenced pastures. Tack shop on premises. Stable specializes in Morgans & driving. Morgan stallion standing at stud: "Beta-B-Protocol." Call for reservations. **Rates:** $15-$20 per night.
Accommodations: Holiday Inn & Sheraton in New Haven, 7 miles away.

GREENWICH
On the Go Farm   Phone: 203-532-4727
Ron Carroll
550 Riverville Road [06831] **Directions:** I-95 to Hwy 684. Call for further directions. Stable is 30 minutes off of I-95. **Facilities:** 20 indoor box stalls, 4 paddocks, 75' x 175' outside arena, easy access to riding trails. English lessons at all levels offered at farm and horse sales. Call for reservations. **Rates:** $22.50 per night. **Accommodations:** Ramada Inn 5 miles from stable.

MILFORD
Spring Meadow Farm   Phone: 203-878-1126
Rhonda M-Alfano   Fax: 203-878-2660
918 Wheelers Farm Road [06460] **Directions:** Call for directions. 2 miles from I-95. **Facilities:** 32 indoor stalls, 80' x 200' indoor arena, large outside riding ring, fenced pasture and separate paddocks. Large heated lounge. English training of horses & riders. Call for reservations. Negitive coggins required. **Rates:** $25 per night. **Accommodations:** Red Roof Inn, Marriot Suites -2 miles away.

DELAWARE

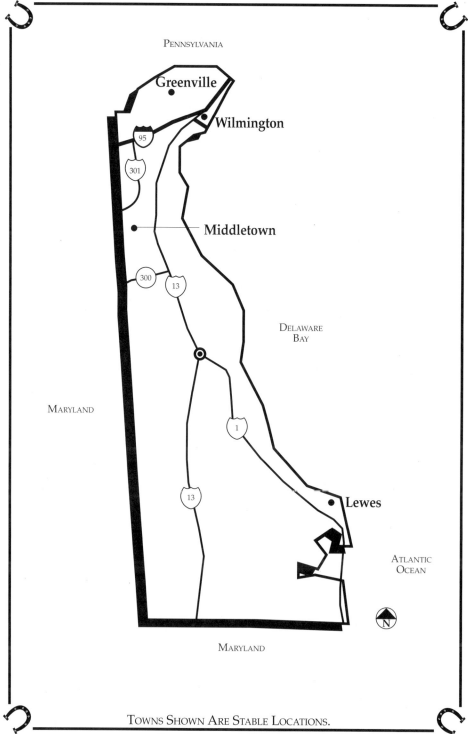

Towns Shown Are Stable Locations.

∗ onsite accommodations

DELAWARE PAGE 57

ALL OF OUR STABLES REQUIRE CURRENT NEG. COGGINS, CURRENT HEALTH PAPERS & OWNERSHIP PAPERS.

GREENVILLE
LRJ Enterprises Phone: 302-655-9601
Laurie Jakubauskas
903 Owl's Nest Road [19807] **Directions:** Delaware Ave. Exit off of I-95. Call for further directions. **Facilities:** 4 indoor stalls, 90' x 180' indoor arena, 50' x 150' outdoor arena, outside jump course, 50 acres of turnout & trails. **Rates:** $25 per night; $140 per week. **Accommodations:** Motels within 5 minutes.

LEWES
Winswept Stables Phone: 302-645-1651
Dawn & Jay Beach
Rt. 24, [19958] Directions: Call for directions. **Facilities:** 12 - 10' x 12' airy stalls, 160' x 200' sand arena, 40 acres, grass jump fields, swimming pond, riding trails, large turnouts and round pen. Located minutes from down town historic Lewes and Rehoboth beaches. Beach riding available. Reservations required. **Rates:** $25 per night, $140 per week. **Accommodations:** Many motels within minutes in Lewes, Rehoboth, & Dewey Beach.

MIDDLETOWN
Price's Public Racing Stable Phone: 302-378-2032
Harry Price
1239 Bunker Hill Road [19709] Directions: Located on Rt. 15 off of I-95. Call for directions. **Facilities:** 4 small and 5 large paddocks, large pasture, and riding trails. Trains and sells race horses. Call for reservations. **Rates:** $20 per night. **Accommodations:** Motels 15 miles away.

WILMINGTON
Twin Pines Farm Phone: 302-478-9917
Doc Talley
5700 Concord Pike [19803] Directions: 5.5 miles north of Wilmington on Rt. 202. Call for directions. **Facilities:** 28 indoor stalls, paddocks, and riding ring. Feed/hay and trailer parking available. Full boarding facility. **Rates:** $20 per night. **Accommodations:** Motels within 1 mile of stable.

PAGE 58 FLORIDA

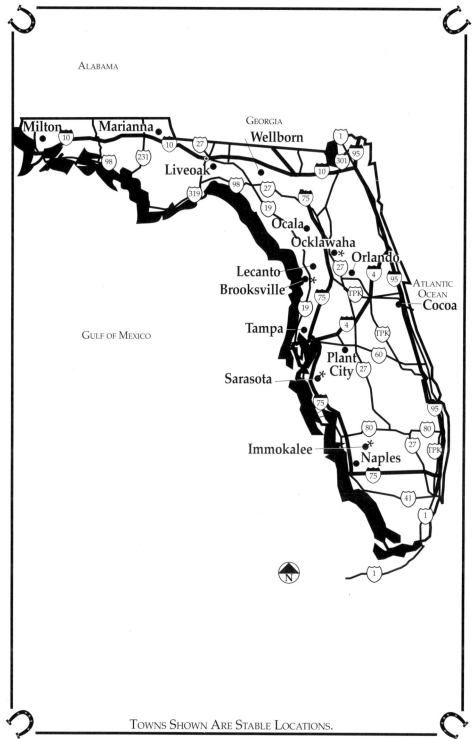

Towns Shown Are Stable Locations.

* onsite accommodations

FLORIDA   Page 59

ALL OF OUR STABLES REQUIRE CURRENT NEG. COGGINS, CURRENT HEALTH PAPERS, & OWNERSHIP PAPERS.

BROOKSVILLE
✱ <u>L & M Paso Fino Ranch</u>   **Phone: 931-879-4944**
Lowell & Melinda Ensinger
5114 Spring Lake Hwy [34601]   **Directions:** From I-75: Exit 61 then west on SR 50. South on Spring Lake Hwy (541). Ranch is on left. **Facilities:** 10 indoor stalls, paddocks & large pastures. 100' x 200' riding ring with lights, round pen, 25 acres with bridle path. L & M Ranch is also a bed & breakfast with heated swimming pool & tennis court. **Rates:** $15 for first night, $10 per night thereafter; $75 per week. **Accommodations:** Bed & Breakfast on site.

COCOA
<u>Ace of Hearts Ranch</u>   **Phone: 321-638-0104**
Dennis & Sandra Bressler   Web: www.aceofheartranch.com
7400 Bridal Path Lane [32927]   **Directions:** From I-95 Exit 79 (Rt #50) East to Rt 405. Turn right. Go to 3rd traffic light, Grissom and turn right. Go 3 miles south past Kings Hwy Light and turn right on Ranch Rd (dirt road). Go 1 mile on right, Bridal Path Lane. Call for U.S.#1 directions. **Facilities:** 14- 12' X 12' rubber mat stalls. Hay and feed available. 1/2 acre turn-out pasture. Trailer parking. Lots of trails. Beach Riding available Nov. - Apr. 5 minutes from Kennedy Space Center. 20 minutes from Cocoa Beach. 45 minutes from Orlando. Please call for reservations. NO STALLIONS. **Rates:** $20 per night, $110 per week. **Accommodations:** Lots of motels -- Titusville 5 miles, Cocoa 8 miles, Cocoa Beach 18 miles, Orlando 45 miles.

IMMOKALEE
✱ <u>Midge Lessor</u>   **Phone: 239-657-4569**
RR 2 Box 560 [34142]   **Directions:** 4 miles south of I-75 on SR 29. **Facilities:** 10 indoor 10' x 12' covered box stalls, lighted regulation-size training arena, feed/hay, trailer parking. Private 160-acre farm, backs up to Fakahatchee Strand State Forest. Can handle stallions with 2 days' notice. **Rates:** $20 per night, discount for more than one ; $50 per week. **Accommodations:** B&B on premises. Motels in Everglades City, 17 miles from stable.

LECANTO
<u>Renab Ranch</u>   Ranch: 352-628-9816
Lillian Baner   Store: 352-628-2716
5338 So. Lecanto Hwy, Rt. 491 [34461] **Directions:** Call for easy directions from I-75 or US 19. **Facilities:** 6 stalls, 21 acres of cross-fenced premium pasture, indoor/outdoor wash down areas, 60' x 90' exercise ring, scenic 1/4-mile bridle path access to 41,000 acres of state forest. Rolling hills and shady trails. Water/elec. hook-up for campers, shower. Reservations required. Owners live on premises and own local feed store; free delivery to your campsite. **Rates:** $10 per night. **Accommodations:** Riverside Inn, Homosassa, 8 miles from ranch.

FLORIDA

ALL OF OUR STABLES REQUIRE CURRENT NEG. COGGINS, CURRENT HEALTH PAPERS & OWNERSHIP PAPERS.

LIVE OAK
Spirit of the Suwanee Music Park Phone: 386-364-1683
3076 95th Drive [32060] **Directions:** 100+ (including portable stall)- minimum size 10'x10'. feed/hay and trailer parking available. This is a large campground (640 acres) that adjoins several thousand acres of water management land, which is great for horseback riding. We have modern barns, a tree house & traditional camping all available on site. In addition, we have a restaurant, canoe rental, horse rental, wedding chapel, 700-seat music hall, craft mall & Suwanee River Frontage. It is best to call ahead to see what is going on in the park. We do have periodic, large outdoor events, which sometimes don't go well with horse activities. **Rates:** $12-$15 per night. **Accommodations:** Best Western, Econolodge, Holiday Express all within 4 to 5 miles.

MARIANNA
Circle D Ranch Phone: 850-352-4882
George E. Dryden
3121 Dryden Drive [32446] **Directions:** From I-10: Exit onto 231 to Cottondale. Turn right onto Hwy 90 at first light. Go 5 miles to ranch entrance on left. **Facilities:** 30 indoor stalls, trailer parking, no feed/hay, no turnout. Must sign a release. **Rates:** $10 per night. **Accommodations:** Days Inn located in Marianna.

MILTON
Coldwater Recreation Area Phone: 850-957-6161
Florida Division of Forestry
David Creamer
11650 Munson Highway [32570] **Directions:** 15 miles north of Milton off Hwy 191. **Facilities:** 69 covered 10' x 12' stalls, 8 outdoor 15' x 15' pens, tie outs for 100+ horses, two 1-acre pasture/turnout areas, trailer parking, feed/hay arranged. Located on Coldwater Creek, Blackwater River State Forest; 35 miles of marked riding trails. Campground, showers, swimming, canoeing. Reservations required. **Rates:** $5per night, $3 per paddock, $13 per campsite. **Accommodations:** Quality Inn in Milton (22 miles), cabins at Adventures Unlimited canoe livery (6 miles).

NAPLES
M & H Stables Phone: 239-455-8764
Rick Morrell, Joyce Holland, & Andreas Kertscher Fax: 239-352-1186
2750 Newman Drive [34114] **Directions:** 2 miles off Exit 15 of I-75. Call for directions. **Facilities:** 51 indoor stalls with connecting paddocks, 200' x 300' riding arena. Hundreds of miles of riding trails, feed/hay & trailer parking on site. Paint stallion standing-at-stud: "Bodacious," a black Tobiano stallion. Training & breaking of horses done at stable as well as Western riding lessons. Call for reservation. **Rates:** $20 per night. **Accommodations:** Comfort Inn, Super 8, Budgetel, and Quality Inn within 3 miles of stable.

FLORIDA   PAGE 61

ALL OF OUR STABLES REQUIRE CURRENT NEG. COGGINS, CURRENT HEALTH PAPERS & OWNERSHIP PAPERS.

OCALA
Dancing Horses Farm of Ocala Phone: 352-873-7084
Robert & Barbara Walla Or: 352-895-7915
Web: dancinghorsesfarm.com E-mail: bwalla@atlantic.net
8711 West Hwy 40 [34482] **Directions:** I-75 Exit 352, West on Hwy 40 for 4.6 miles driveway on right. **Facilities:** 18 stalls. Matts-CB-12 X12, bug spray, auto water. Trailer parking available. Dressage arena, training facility, riding trails. **Rates:** $25 per night. **Accommodations:** Many major hotels. RV hook up available.

OCKLAWAHA
* **Wit's End Farm** Phone: 352-288-4924
Margo Atwood-Langstaff Fax: 352-288-8147
P.O. Box 964, [32183] **Directions:** 20 miles east of I-75 & 10 miles from SR 40. Call for further directions. **Facilities:** eight 12' x 12' indoor stalls, 1/2-acre areas of turnout, feed/hay. Trailer parking and limited RV parking available. Limited smoking. **Rates:** $10 per night. **Accommodations:** Lakefront 2 Bedroom rental apartment. Swimming and canoeing. Other motels in Ocala, 20 miles away. Close to Ocala National Forest.

ORLANDO
Grand Cypress Equestrian Center Phone: 800-835-7377
Liz Trellue or 407-239-1938
One Equestrian Drive [32836] **Directions:** Call for directions. **Facilities:** 44 stall barn with 10 used for overnight travelers, 8 turnout paddocks, 24 hr security, automatic fly spray & watering systems, shavings for bedding, feed/hay, trailer parking. Center has received the British Horse Society's "Approval" for livery and training; instruction available, show series, clinics. Reservations required. **Rates:** $35 per night includes feeding. **Accommodations:** The Villas of Grand Cypress and Hyatt Regency Grand Cypress on same property.

PLANT CITY
Turkey Creek Stables Phone: 813-737-1312
Peggy Womack
5534 S. Turkey Creek Road [33567] **Directions:** Exit 10 off of I-4 to Turkey Creek, south of State Road 60. **Facilities:** 10 indoor stalls, 3,000 acres of pasture, turnout, and trails. **Rates:** $8.00 plus .50 per feeding; weekly rate available. **Accommodations:** Holiday Inn & Best Western, Plant City, 7 miles from stable.

PAGE 62 FLORIDA

ALL OF OUR STABLES REQUIRE CURRENT NEG. COGGINS, CURRENT HEALTH PAPERS, & OWNERSHIP PAPERS.

SARASOTA

Farr Crest Farms　　　　　　　　　　　Phone: 941-322-9519
Paige Farr　　　　　　　　　　　　　　　　Cell: 941-232-4029
15910 Rawls Rd [34240]　　　　E-mail: farrcrest@mailmt.com
Directions: I-75 exit 210, go east on Fruitville Rd to end; turn right onto Myakka Rd, go 1-1/4 miles, turn right onto Rawls Rd, 2nd driveway on left. **Facilities:** 8 12'x12' indoor stalls, 2 paddocks and 5 acre pasture. New Barnmaster barn 60ft round pen, grass arena, barrel area, 2 wash racks, and foaling stalls available. Hay is provided; you provided your own grain or sweet feed. Trailer parking available. **Rates:** $20 per night, $120 per week. **Accommodations:** AmericInn 10 min at I-75 exit 210 Sarasota.

✶ **Willoughby Farms**　　　　　　　　　Phone: 941-379-5220
Larry Friedman　　　　　　　　Web: www.willoughbyfarms.com
1201 Sinclair Drive [34240]　**Directions:** Exit 39 off rt. 75, East 4 miles to Sinclair Drive, turn left follow to Willoughby sign. **Facilities:** 3 12x12 0r 10x12 stalls, feed/hay available, trailer parking, 3and 5 acre pastures, 100 ft. round pens, exercise arena, optional trail rides, breeding farm for Haflingers, Standing Stud Grand Champion Means Magic "Amadeus". No Stallions Please **Rates:** $20 per night. **Accommodations:** Bed & Breakfast, Full 3-room apartment available.

WELLBORN

Imperial Oaks Ranch　　　　　　　　Phone: 386-963-2908
Jerry & Bobbi Fenderson
16648 53rd Road [32094] **Directions:** I-72 to exit 92/Lake City, west on rt. 90 1/4 mile then left on Pinemount Rd (Rt. 252) 9 miles to left on 53rd Road, 1 mile on left. From I-10 take exit 41, south on 137 to rt. 252, west on 252 for 1.9 miles to left on 53rd road, 1 mile on left. **Facilities:** 4 inside 12'x12' stalls, 8 outdoor stalls, 2/ large run-ins. 2 round pens, many trails. 2 pastures. Hay/feed available. Trailer parking, self contained camper trailers and 18 wheelers O.K. Breeders of Morgan Horses, Standing Paramout Imperial@ Stud. Call ahead for reservations. **Rates:** $15 per night.

Wellborn Quarter Horses　　　　　　Phone: 386-963-1555
Andrea & Joe Schomburg　　　　　　　　　Fax: 386-963-1557
　　　　　　　　　　　　　　　　　　　　Cell : 386-623-1099
8660 CR-137 **Directions:** Exit 292 off I-10, 1/2 mile north on 137. Exit 439 off of I-75. Call for Directions. **Facilities:** 4 12x12 indoor and 2 12x12 outdoor stalls, Round pen, 11 pastures on 18 acres, Feed/Hay available, Trailer parking, Stallion-Son of Impressive (HYPP/NN) standing, Quality performance horses occasionally for sale, dogs on leash, stallions okay, semis okay. **Rates:** $20 per night, $100 per week. **Accommodations:** Motels off exit 439 on I-75, McLeran B&B in Wellborn.

NOTES AND REMINDERS

Page 64 GEORGIA

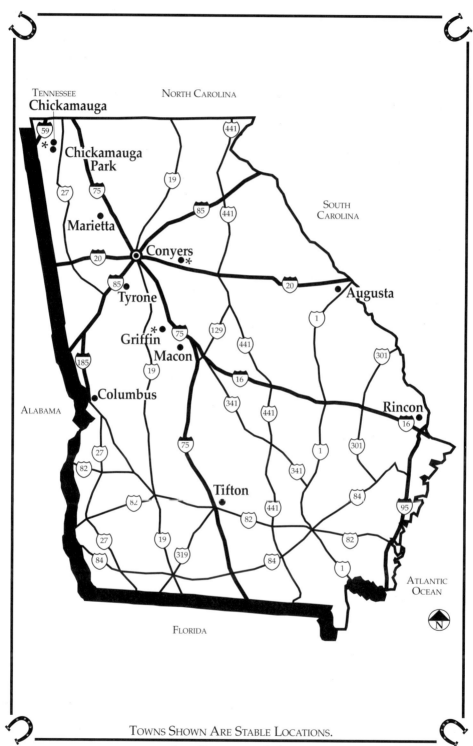

Towns Shown Are Stable Locations.

* onsite accommodations

GEORGIA PAGE 65

ALL OF OUR STABLES REQUIRE CURRENT NEG. COGGINS, CURRENT HEALTH PAPERS & OWNERSHIP PAPERS.

AUGUSTA
Augusta Riding Center Phone: 706-863-9044
Ruth & Jim Jatho
1403 Flowing Wells Road [30909] **Directions:** From I-20: Bel-Air Road exit. Turn south & take first road on left (Frontage Rd.). Go down 1 mile & turn right on Flowing Wells Road. Center is 1/2 - 3/4 mile on right. **Facilities:** At least 5 indoor stalls, 3 large pastures & 3 paddock areas, feed/hay at extra cost, trailer parking available. Call ahead for availability. **Rates:** $20 per night. **Accommodations:** Ramada Inn & EconoLodge 2 minutes away.

CHICKAMAUGA
* **Chickamauga Bed & Breakfast & Stable** Phone: 706-375-3476
Kay Red Horse
P.O. Box 81 [30307] **Directions:** 20 minutes SE of Chattanooga, Tennessee. Call for directions. **Facilities:** 5-stall barn, pasture turnout, trailer parking. Across the road from the Civil War Battlefield with 12 miles of trail riding. Raise and show spotted Tennessee Walkers. Nonsmoking, no alcohol. Reservations required. **Rates:** $15 per horse. **Accommodations:** Bed & Breakfast on 14-acre farm. Full country breakfast; other meals on request. $55 per couple.

CHICKAMAUGA PARK
Trail's End Ranch Riding Stables Phone: 706-375-4346
Sarah Clinton
Hwy 27 South, Trail End Road [30707] **Directions:** 18 miles from Chattanooga. Call for directions. **Facilities:** 5 indoor stalls, pasture/turnout available, feed/hay for purchase, trailer parking. 3-7 days notice. **Rates:** $10 w/o feed. **Accommodations:** Best Western 3 miles from stable.

COLUMBUS
Shamrock Stables Phone: 706-561-9103
Elvin Amon
6620 Moon Circle [31909] **Directions:** From I-185: Take Exit 6 (Airport Thruway), go east 4 miles to Moon Circle on right. **Facilities:** 3 stalls in a pole shelter, paddocks available, feed/hay & trailer parking available. One day notice if possible. **Rates:** $10 w/o feed. **Accommodations:** Several nearby.

CONYERS
* **Linda's Riding School** Phone: 770-922-0184
Linda Greene Ridley Web: lindasridingschool.com
3475 Daniels Bridge Road [30094] **Directions:** Take Exit 74 off of I-20E. Call for further directions. **Facilities:** 10 indoor stalls, holding pen, riding rings and trails. Hay & trailer parking available. Please call for reservations. **Rates:** $20 per night, $40 if arrival after 6 P.M. **Accommodations:** Have cots above barn and apartment for rent. Motels within 8 miles of stable.

Page 66 GEORGIA

ALL OF OUR STABLES REQUIRE CURRENT NEG. COGGINS, CURRENT HEALTH PAPERS, & OWNERSHIP PAPERS.

GRIFFIN
Silver Horseshoes Stables, Ltd.   Phone: 770-227-7681
Lisa Goldman, Manager   or: 770-714-7681
E-mail: myhank@aol.com   Web: www.silverhorseshoes.com
2946 High Falls Road [30223] **Directions:** I-75 to Exit 205, go west, 5 miles to caution light, turn right onto High Falls Road. Stable is 6th driveway on left. **Facilities:** 30 indoor 14' x 14' box stalls, pasture, safe fence, indoor and outdoor arenas, feed/hay available, trailer parking, professional management. Clean, modern full care facility, Riding lessons, boarding. **Rates:** All stalls $25 before 5 PM, $40 after 5 PM. **Accommodations:** Comfort Inn, Days Inn, Holiday Inn, Hampton Inn within 10-12 miles of stable.

GRIFFIN/MILNER
✶ Pegasus Riding School, Inc.   Phone: 770-228-3865
Linda & Warren Abrams   E-mail: pegride@bellsouth.net
Web: www.pegride2000.com
392 Philip Weldon Rd, Milner (30257) **Directions:** Southbound on I-75 Exit 201; Northbound Exit 198. Call for directions. **Facilities:** 17 indoor, shavings included, 12x12 stalls, 7 indoor 10x10 stalls, 5 acre pasture, turnout, tack shop, feed/hay available, private paddocks, space for large rigs. RV hookup. Hunter/jumper and dressage riding. Small pets welcomed $5 per pet, large pets no charge in barn. **Rates:** $20 per night, $125 per week. **Accommodations:** Apartment sleeps up to 6 people $100 per night or $25 per person, shower, A/C, TV, VCR, kitchen, micro, refrig, coffee maker. A smoke free facility, cash only.

HOGANSVILLE
Flat Creek Ranch, Inc.   Phone: 706-637-8920
Joan Keegan, Manager
3564 Mountville Rd. [30230] **Directions:** I-85 Exit 6. West towards Hogansville. Follow campground signs. Immediate left at Waffle House (Bess Cross Road). 2 miles to 4-way stop. Left onto Mountville - Hogansville Road. Campground on right. **Facilities:** 70 - 10' X 14' indoor stalls. No feed or hay available. No pasture or turnout available. Trailer parking space provided. **Rates:** $10 per night, $200 per month. **Accommodations:** Flat Creek Campground 706-637-6001. Key West Motel 3 miles 706-637-9395. Hummingbird Inn - 3 miles 706-637-5400

MACON
Wesleyan College Equestrian Center   Phone: 478-757-5103
Jon Conyers, Director of Riding   E-mail: JConyers@WesleyanCollege.edu
4760 Forsyth Road [31210] **Directions:** Exit #3 (Zebulon Road) off I-475. See signs for Wesleyan College. **Facilities:** New facility completed summer 1999, 24 indoor stalls, 5 outside paddocks, lighted outdoor riding arena, 40 acres of wooded trails. Location of Collegiate Equestrian Team and Community Horsemanship Program. **Rates:** $30 per night. **Accommodations:** Jamison Inn (912-474-8004) and Fairfield Inn (912-474-9922) 2 miles from stable.

GEORGIA   Page 67

ALL OF OUR STABLES REQUIRE CURRENT NEG. COGGINS, CURRENT HEALTH PAPERS, & OWNERSHIP PAPERS.

MARIETTA
Hymnbrook Farm   Phone: 770-428-1065
Barbara & Bill Dawson
150 Mt. Calvary Road [30064] **Directions:** Location is 25 miles North of Atlanta, GA. Farm is 5 miles from I-75 and there are 7 motels at exit 269. From I-75, take Exit 269 going West. Cross Hwy 41, Hwy Old 41, Stilesboro Rd and at Burnt Hickory Rd, turn left onto Burnt Hickory and immediate right to Mt. Calvary Rd. (about 200'). Farm is about 1 mile on right. 150 Mt. Calvary Rd. **Facilities:** 10 12 x 12' indoor stalls, ring, trailer parking and electric available. Owner on premises. **Rates:** $20 per night including bedding. **Accommodations:** Six motels within 5 miles.

RINCON (Savannah area)
Hi Ho Hills Farm   Phone: 912-826-5808
Robin Hughes
939 Goshen Road [31326] **Directions:** Exit 19 off of I-95 to Hwy 21 N. 4.5 miles to Goshen Road. Take left. Farm is 1.5 miles on left. **Facilities:** 12 indoor stalls, round pen, 3 large fields, show & dressage rings, 20 miles of trails, & 2 camper spaces available. Feed/hay & trailer parking available. 24-hr notice required. **Rates:** $15 per day; $100 per week. **Accommodations:** Sleep Inn, 5 miles away on Hwy 21 & I-95.

TIFTON
Windy Hill Stables   Phone: 229-386-8108
Gary & Mary Anne Simmons   Fax: 229-686-7187
4186 Hwy 82 [31794] **Directions:** Exit 18 off I-75 4.5 miles west on 82. White fence and big red barn. Number on mailbox. **Facilities:** New barn, steel, kickproof, Barn Master 12- 12 X 12 indoor stalls. 1 acre pasture turnout and round pen. 50 acres with lots of dirt roads to ride. 800 acres permission to ride. Golf course 1.5 miles away. **Rates:** $15 per night. **Accommodations:** 4 miles from 8 motels.

TYRONE
Trickum Creek Ranch   Phone: 770-487-2146
Debbie Lowe   Fax: 770-486-1075
Web: www.debbielowe.com   E-mail: horsehotel@juno.com
213 Lincoln Road [30290] **Directions:** I-85 South to exit 56. Turn East, go exactly 2.2 miles, turn left at Lincoln Road, half mile to #213. **Facilities:** 8 12x12, 10x10 indoor stalls, feed/hay available, trailer parking, 3 multi-acre pastures, 2 acre runs with stalls, Studs welcome. **Rates:** $20 per night. **Accommodations:** Holiday Inn, Days Inn, Hampton Inn, Jamison Inn, Motel 6 and more less than 5 miles.

PAGE 68   **IDAHO**

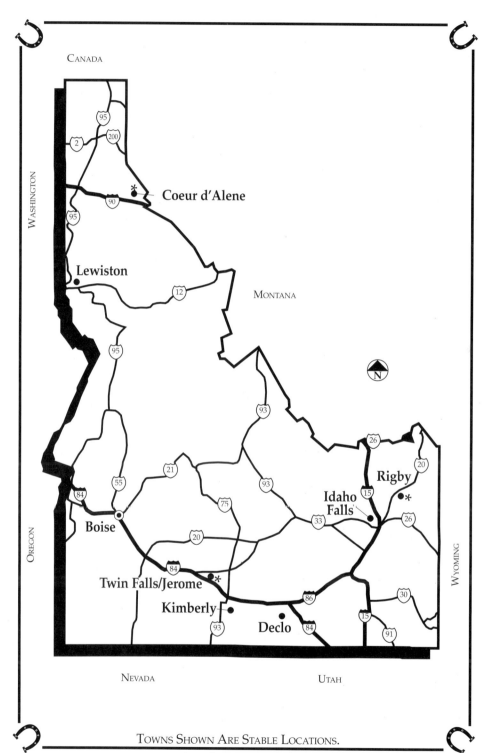

Towns Shown Are Stable Locations.

* onsite accommodations

IDAHO Page 69

ALL OF OUR STABLES REQUIRE CURRENT NEG. COGGINS, CURRENT HEALTH PAPERS, & OWNERSHIP PAPERS.

BOISE/KUNA/MERIDIAN
Aspenbreak Stables Phone: 208-922-4563
John & Karen Vehlow
E-mail: jkvehlow@overarch.com   Web: www.hi-sagemorgan.com
330 N. Eagle Road [83634] **Directions**: Exit 44 off of I-84 (about 7 miles west of Boise). Go south approx. 6 miles, turn left on Kuna Rd. 2 miles to Eagle Rd., take left onto Eagle Rd., go about 300 yards on Eagle Rd. to Flying Horse Lane. Go left on Flying Horse Lane. Arena and stalls are in first building on right side of road. **Facilities:** 10 12x12 indoor stalls and pens available for turnouts, 90' x 50' pasture/turnout, large indoor and outdoor arenas, & riding trails. Vet/farrier on call. **Rate:** $15 per night. **Accommodations:** Available at Exit 44.

BOISE
Hidden Valley Ranch-Paint Horses Phone: 208-362-4345
Sabrina Leonard Fax: 208-362-4363
Web: hiddenvalleypaints.com E-mail: brina@overarch.com
8699 S. Gantz Ave [83709] **Directions:** I-84, Call for Directions. **Facilities:** 3 12x14 stalls with big runs, feed/hay available, trailer parking, several huge corrals, arenas, pasture, wash bays, Skip Y2K2 -Paint Stallion. **Rates:** $15 per night. **Accommodations:** Hampton Inn off I-84, 5 miles away.

COEUR d'ALENE
* **Kingston 5 Ranch Bed & Breakfast** Phone (local): 208-682-4862
Walter & Pat Gentry Toll Free: 1-800-254-1852
Web : www.k5ranch.com Fax: 208-682-9445
P.O. Box 2229, Coeur d'Alene [83816] **Directions:** Take I-90 to Exit 43 South to Frontage Road (1-2 blocks), turn west. Ranch is exactly 1 mile on right. 2 hrs west of Missoula, Montana & 1 hr east of Spokane, Washington. **Facilities:** 2 indoor/outdoor stalls, 4 individual pasture areas, all areas with panels or safe 2x4 horse fence. Also, round pen with 2 attached outdoor pens, 200' x 100' arena, parking, and famous 1,000-mile Silver Country Trail System. **Rates and Accommodations:** Relax and pamper yourself in one of our Jacuzzi/Fireplace suites with in-room private bath & Jacuzzi tub, in-room fireplace and personal outdoor hot tub for each room. Rates start @ $99.50 per night + 7% tax for 1or 2 guests and 1 horse. Each additional horse $10.00. Veterinarian services available. No smoking or pets indoors. Other outside pets by prior arrangement only. Overnight horse accommodations only available with B&B stay.

DECLO
Milo Erekson Phone: 208-654-2085
E-mail: miloerekson@hotmail.com
200 North Highway 77 [83323[**Directions:** 1/4 mile south of I-84 Exit 216
Facility: Up to 10-acre pasture only, feed/hay available, trailer parking. **Rates:** $10 per night, $50 per. week **Accommodations:** Burley has several motels 10 miles from stable.

IDAHO

ALL OF OUR STABLES REQUIRE CURRENT NEG. COGGINS, CURRENT HEALTH PAPERS, & OWNERSHIP PAPERS.

IDAHO FALLS
C & D Stables   Phone: 208-522-1439
Donna Garriott
3909 North 15 E. [83401] **Directions:** I-15, Exit 119 to US 20, 2 miles to Exit 15 E. Stable 1/4 mile. **Facilities:** 24 - 10' x 12' stalls, heated watering units, indoor and outdoor arenas, alfalfa/oats available, trailer parking. **Rates:** $15 per night. **Accommodations:** Motels 2 miles away in Idaho Falls.

Ellis Supreme Arabians   Phone: 208-524-7247
Terie Ellis   E-mail: tmellis@if.rmci.net
1438 West 97 South [83402] **Directions:** Coming from North on I-15. Exit 113, turn left. Go straight through stop light, across railroad tracks to next stop sign. Make a right, go 2 miles. We are on the North East corner. Coming from South on I-15. Exit 113, turn right. **Facilities:** 27 indoor 12' x 12' stalls, 14 with outside runs, 70' x 150' indoor arena, 160' x 160' outdoor arena, round pen, grooming stalls, year-round wash bay, heated tack room, & breeding laboratory. Sales & training of horses. Call for reservations. **Rates:** $15 per night; feed and hay $5 extra. **Accommodations:** Evergreen Gables Motel 5 miles from stable, Shilo Best Western 7 miles from stable.

KIMBERLY
Western Barns   Phone: 208-423-6340
Carol L. Sherman   E-mail: csherman@onewest.net   208-731-3557
3216 East, 3625 North [83341] **Directions:** I-84 to Twin Falls exit 182, South West on Hwy 50/30, left onto Hankins Rd, left at end, second place on left. Large white barn with green roof. **Facilities:** 8- 12' x 12' box stalls, 6-24' x x24' outdoor pipe corrals with shelter, 80' x 100' pasture/turnout. No feed/hay available. Horse must be de-wormed 1 month prior. Modern, clean & safe facility. Guide available for mountain trails. **Rates:** Call for rates. **Accommodations:** Motels in area.

LEWISTON
Lewiston Roundup Association   Phone: 208-746-6324
7000 Tammany Creek Road [83501]   Phone: 208-746-7589
Directions: Take Hwy 12 to 21st Avenue. Call for further directions. **Facilities:** 122 indoor stalls, 150' x 240' outdoor arena, 200' x 300' indoor arena, trailer parking & RV Park with electric, water, & sewer hook-ups for 60 vehicles. Open 6 A.M. - 10 P.M. with year-round groundskeeper. Rodeos held on 80-acre facility. Call for reservations. **Rates:** $15 per night and must clean stall before leaving. **Accommodations:** Grand Plaza Hotel 5 miles from stable (208-799-1000).

IDAHO   Page 71

ALL OF OUR STABLES REQUIRE CURRENT NEG. COGGINS, CURRENT HEALTH PAPERS, & OWNERSHIP PAPERS.

RIGBY
✱ <u>Blacksmith Inn</u>   Phone: 208-745-6208
   Mike & Karla Black   Web: www.blacksmithinn.com
227 North 3900 East [83442] **Directions:** Call for directions. **Facilities:** 1 indoor 16' x 16' stall, 6 outdoor 16' x 16' stalls, 2 corrals, turnout, feed/hay available, trailer parking available. Raise and breed Tennessee Walkers, **Rates:** $10 per night with reservations. Weekly rate negotiable. **Accommodations:** Bed & Breakfast on premises.

TWIN FALLS/JEROME
✱ <u>Days Inn</u>   Phone: 208-324-6400
   Web: www.daysinn.com/jerome13000
   E-mail: daysinn3000@prodigy.net
1200 Centennial Spur [83338] **Directions:** Exit 173 off I-84 1/2 mile north next to Flying E Truck stop on Hwy 93. **Facilities:** 20 10' x 10' outdoor stalls. Trailer parking available. **Rates:** $12 per night. **Accommodations:** Days Inn, stables located on property.

Page 72 ILLINOIS

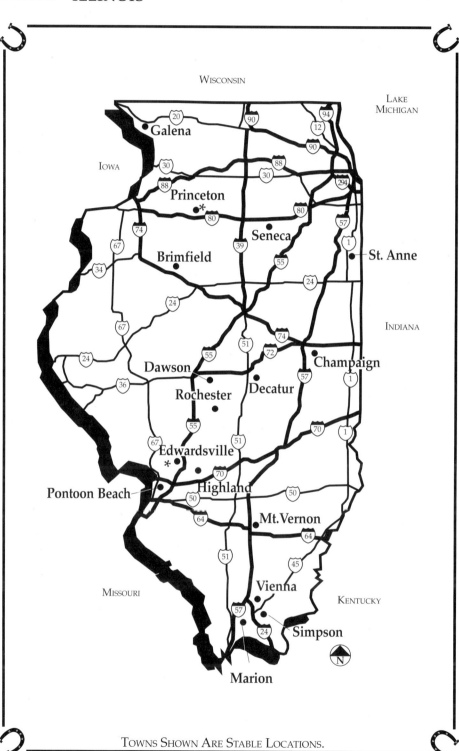

Towns Shown Are Stable Locations.

* onsite accommodations

ILLINOIS   Page 73

*ALL OF OUR STABLES REQUIRE CURRENT NEG. COGGINS,
CURRENT HEALTH PAPERS, & OWNERSHIP PAPERS.*

BRIMFIELD (KICKAPOO)
Rocky C Stables   Phone: 309-446-9306
Jenni & Ean Cuthbert
6628 N. Kramm Road [61517]   **Directions:** 1.5 miles from I-74 at Exit 82. Call for directions. **Facilities:** 10 indoor stalls, 3 - 100' x 100' turnout paddocks, indoor arena, wash rack, feed/hay available. Standing-at-stud: "Charlie's Out of Oil." Call for reservations. **Rates:** $20 per night. **Accommodations:** Holiday Inn in Peoria, 15 minutes away.

CHAMPAIGN
Unzicker Stables   Phone: 217-359-5641
Carolyn Unzicker
1162 County Road 900E [61822]   **Directions:** 1.5 miles east of I-57 at Exit 229. Call for directions. **Facilities:** 40 indoor stalls, large indoor & outdoor riding arenas, & tack store on premises. Call for reservations and availability. **Rates:** $15 per night; $75 per week. **Accommodations:** Best Western Paradise Inn 3 miles away and LaQuinta 5 miles from stable.

DAWSON
Scofflaw Farm   Phone: 217-364-4350
Donald R. Lawler
RR 1, Box 93A [62520]   **Directions:** 8 miles from I-55 & 6 miles from I-72. Call for directions. **Facilities:** "One of the finest horse facilities in Central Illinois." 8 indoor box stalls, 11 large turnout paddocks with lean-tos, two 36' x 40' holding pens, 80' x 160' indoor arena, 200' x 247' outdoor ring with large gazebo. No stallions. Call for reservations. **Rates:** $20 per night; discounted weekly rate. **Accommodations:** Holiday Inn & Days Inn in Springfield, 8 miles away.

DECATUR
Skyline Stables   Phone: 217-422-1051
Linda & Ed Seaton   Barn: 217-422-7630
4095 Rock Springs Road [62522]
Directions: 3 miles from I-72 & 1/2 mile from both Hwy 48 & 51. Call for directions. **Facilities:** 4 indoor stalls, 1/4- to 6-acre paddocks, indoor & outdoor arenas, feed/hay & trailer parking available. Breeds Arabians. Stallion services available. Call for reservations. **Rates:** $15 per night; discounted rates long term. **Accommodations:** Holiday Inn in Decatur, 3 miles from stable.

EDWARDSVILLE
* The Sleepy P   Phone: 618-659-1051
John & Victoria Piel   Web: www.sleepy-p.com   E-mail: sleepyp@spiff.net
6309 Miller Drive, Edwardsville, IL [62025]   **Directions:** Call for directions. **Facilities:** 10 Stalls and pastures/turnout. All amenities to comfortably accommodate your equine traveler. Feed/hay and vet/farrier on call. **Rates:** Call for rates. **Accommodations:** Bed (comfortable accommodations). Camper/RV hookup and primitive camping. Scenic area. Excellent Indian Museum nearby.

ILLINOIS

ALL OF OUR STABLES REQUIRE CURRENT NEG. COGGINS, CURRENT HEALTH PAPERS, & OWNERSHIP PAPERS.

GALENA
Shenandoah Riding Center Phone: 815-777-2373
Galena Territory Association Fax: 815-777-0581
Susan Miller Web: www.shenandoahridingcenter.com
E-mail: src@galenalink.com
200 N. Brodrecht Road [61036] **Directions:** From I-94 near Rockford, go west on Hwy 20 to entrance of Galena Territory. **Facilities:** 50 indoor stalls, lighted indoor arena, 2 outdoor arenas, cross country jumps, club room with viewing area, wash stall, riding lessons, and trails, varying sizes of pasture/turnout, feed/hay & trailer parking available. **Rates:** $35 per night. **Accommodations:** Eagle Ridge Resort, 3 miles, Ambercreek Rentals, 3 miles. Numerous Hotels, Motels and B&B's are available in Galena, 6 miles.

HIGHLAND
Highland Horse Stables Phone: 618-654-3401
Kevin & Patti Gleason Cell: 618-567-5477
E-Mail: kgleasonhotdj@hotmail.com
12118 Ellis Road [62249] **Directions:** Close to major highways. Call for directions. **Facilities:** 70 – 10X10 indoor stalls. Feed/hay available, trailer parking available. 5 acre pasture, fenced and cross-fenced, 60 X 150 indoor arena, 110 X 220 outdoor arena. **Rates:** $20 per horse per night. **Accommodations:** Holiday Inn Express, Cardinal Inn- 3 miles from stable.

MARION
WildHorse Stables Phone: 618-964-1165
Rich & Brenda Rybak E-Mail: wildhors@midwest.net
5094 South Market Road [62959] **Directions:** Call or E-mail for directions. **Facilities:** 6 10x20 and 4 10x10 rubber matted stalls, feed/hay available, trailer parking, various size turnouts, vet stall, heated and air conditioned overnight rooms with bathroom and shower. **Rates:** $25 stall, $15 unsheltered pens, per night, per horse. **Accommodations:** Holiday Inn Express, Comfort Inn, Comfort Suites, Drury Inn, and others nearby.

MT. VERNON
Richardson Stables Phone: 618-242-6566
C. Wayne & Judy Richardson Stable: 618-242-1232
11246 N. General Tire Lane [62864] **Directions:** 1 mile east of I-57 on I-64 to Exit 80. Turn north on IL Rt. 37, 1.8 miles to first stop light. Turn right on IL Rt. 142, 1.5 miles to Richardson's sign. Turn right & stable is on the left. **Facilities:** 14 indoor stalls, one 3-acre pasture/turnout lot, trailer parking, & camper hook-up. Reservations requested. **Rates:** $15 per night. **Accommodations:** Many motels & restaurant in Mt. Vernon, 3 miles from stable.

ILLINOIS Page 75

ALL OF OUR STABLES REQUIRE CURRENT NEG. COGGINS, CURRENT HEALTH PAPERS, & OWNERSHIP PAPERS.

PONTOON BEACH
<u>Gateway Stables</u>   Phone: 618-931-3527
Kelly Arnold   E-mail gatewaystables@sbcglobal.net
3514 Lake Drive [62040]   **Directions:** Just north of Horseshoe Lake State Park. One block off Hwy 111. Convenient to Interstates 270, 70, 55, 64, 255. Call for directions. Minutes from St. Louis & Fairmount Park (T.B. Racing). Resident farrier and good vets on call. Please call day of arrival to confirm. **Facilities:** 24 stalls, all sizes of stalls from mini to 16' x 16', turnout paddocks, 1- and 3-acre pastures, 1-acre dry lot, outdoor arena, 60' x 120' indoor arena, feed/hay at additional charge. **Rates:** $20 per horse. **Accommodations:** Ramada Limited, Super 8 in Pontoon Beach, Best Western in Mitchell, smaller local motels available with reservations.

PRINCETON
* <u>The Prairie Hill - Barn, Bed & Breakfast</u>   Phone: 815-447-2487
Janine Klayman, Owner; Janet, Manager   Fax: 815-447-2428
E-mail: klaymanj@aol.com
25457 1275 North Ave. [61356]   **Directions:** 8 miles from I-80. Call for directions. **Facilities:** Country estate & horse facility. 5 - 10' x 10' indoor stalls, feed/hay, & bedding. 5 paddocks, available weather allowing, all post and rail fencing. Riding school and combined training facility. Facility offers 30 acre cross-country course, 2 arenas, miles & miles of trails. Reservations in advance. Health certificate and negative coggins required. **Rates:** $25 per horse per night. **Accommodations:** Historic, antique filled farmhouse with 6 bedroom B & B including homemade breakfast, hot or on the go. Great vacation destination. All pets welcome. Stabling for horses of guests only.

ROCHESTER
<u>Willow Creek Farms</u>   Phone: 217-498-8136
Nancy Wright   or 217-498-9859
4818 Oak Hill Road [62563]   **Directions:** 4 miles from I-55. 6 miles from I-72. Call for directions. **Facilities:** 2 indoor stalls, 2 - 150' x 150' paddocks, 3-5 acres of pasture. Feed/hay and trailer parking available. Hunter/jumper facility. Call for reservations. **Rates:** $15 per night. **Accommodations:** Red Roof Inn & Days Inn in Springfield, 2.5 - 3 miles away.

ST. ANNE
<u>Sunrise Farms, Inc.</u>   Phone: 815-932-6170
Karen Hemza, Manager   815-935-8897
4370 E. 3500 South Road [60964]   **Directions:** 15 minutes east of I-57. Call for directions. **Facilities:** 10 indoor stalls, ten 1-acre paddocks, 4 with run-in shelters, indoor & outdoor arenas, stadium & cross-country jumps. Bring own buckets. Call for reservations. **Rates:** $10-$20 per night. **Accommodations:** Motel 8 & Howard Johnson's in Bourbonnais, 15 minutes away.

ILLINOIS

ALL OF OUR STABLES REQUIRE CURRENT NEG. COGGINS, CURRENT HEALTH PAPERS, & OWNERSHIP PAPERS.

SENECA
The Pony Place　　　　　　　　　　　　　　　Phone: 815-695-5913
Elaine Owens
2945 N. 37th [61360] **Directions:** Adjacent to I-55 and I-80. Call for directions (let ring 10-15 times). **Facilities:** 8 indoor 12' x 12' boxes in new barn with concrete floor, feed/hay, trailer parking, limited turnout, dry lot available. Traveling horses will be stalled at night. Specialize in large children's riding ponies. Welsh Cob at stud: "Winks Titus"; Black Morgan at stud: "CassaNovas' Cashon." Welsh Cobs & Morgan Crossbreeds for sale. **Rates:** $20 per night. **Accommodations:** Local campground nearby; onsite camping, no hook-ups. Motels within 15 miles in Ottawa and Morris.

SIMPSON
Livin' Color Farm, Inc.　　　　　　　　　　　Phone: 618-695-3570
Dennis& Terri Marr   E-mail: livincolr@aol.com   Fax: 618-695-3571
RR 1, Box 221A [62985]   **Directions:** Call for directions. **Facilities:** 12 indoor 10' x 12' stalls, individual turnout pens, 10+ acres pasture, indoor & outdoor arenas, round pen, wash rack, feed/hay available, trailer parking. Extra services (lunging, grooming, etc.) provided for reasonable fee. Retirement home for horses, lay-ups welcome. Call for reservations or e-mail. **Rates:** $15 per night; $75 per week. **Accommodations:** Motels within 10 miles at Vienna.

VIENNA
Shilo Farms　　　　　　　　　　　　　　　　Phone: 618-995-9443
Dee Dee Adams
1290 Bowman Buttons Road [62995] **Directions:** From I-57: Take Exit 40 1.5 miles east. Turn south on gravel road at sign saying 175E & 1825N. Second house on left. Call for directions from I-24. **Facilities:** 3 outdoor covered stalls, 100' x 200' riding arena, feed/hay & horse trailer parking. 90,000 acres of horse trails in national forest & state park within 1 mile. Reservations required. **Rates:** $15 per night; $10 second night. **Accommodations:** Toupal's Country Corner 8 miles away.

NOTES AND REMINDERS

INDIANA

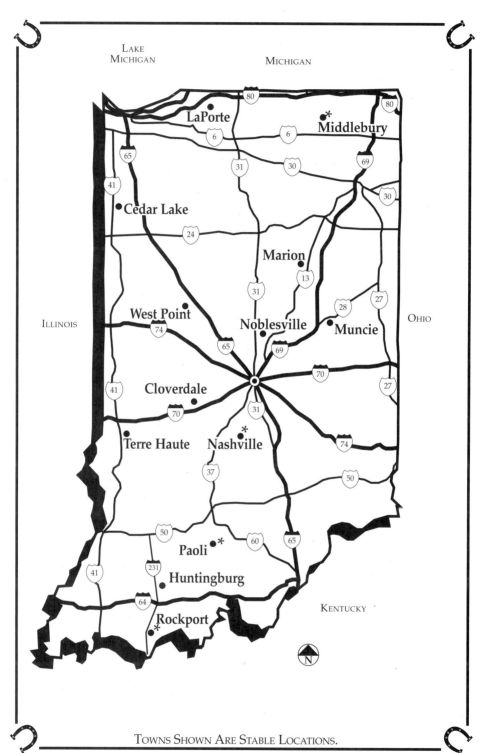

Towns Shown Are Stable Locations.

* onsite accommodations

INDIANA Page 79

ALL OF OUR STABLES REQUIRE CURRENT NEG. COGGINS, CURRENT HEALTH PAPERS, & OWNERSHIP PAPERS.

CEDAR LAKE
Rainbow Equestrian Center, Inc. Phone: 219-365-2127
Ilse Vogelmann
11750 W. 117th Avenue [P.O. Box 968] [46303] **Directions:** 4 miles from Rt. 30; 1/2 mile from Rt. 41; 7 miles from I-65; and 8 miles from I-80 & I-294. Call for directions. **Facilities:** 5 indoor stalls, many size areas of pasture/turnout, feed/hay & trailer parking available. Call for reservation. **Rates:** $25 per night; $20 for 5 or more nights. **Accommodations:** Crestview Motel in Cedar Lake, 1 mile from stable.

CLOVERDALE
Shaky Hill Ranch Horse Motel Phone: 765-795-6344
Kathy Jordan Cell: 765-721-3378
E-mail: ranch@ccrtc.com
7837 S US Hwy 231 [46120] **Directions:** Interstate 70 exit 41 (Hwy 231) go north 1 mile. Stable on the east side of the road. Sign on mailbox. **Facilities:** 6 indoor 10'x12' stalls. Hay/feed available, trailer parking. Half acre of pasture and turnout. **Rates:** $15 per night. **Accommodations:** Super 8 Motel, Motel 6, Ramada Inn, Holiday Inn Express, Days Inn all in Cloverdale 1 mile from stable.

HUNTINGBURG
Willow Creek Ranch, LLC Phone: 812-683-2877
Jud Collet Cell: 812-630-2660
E-mail: willowcreek@psci.net
7021 W. St. Rd 64 [47542] **Directions:** I-64, take exit 54 north on to 161, go straight 7 miles to old 64. **Facilities:** 6-12X9 indoor stalls with electric waterers. 15 acres pasture, paddock area. Also available for transporters to drop off and pick up. Vet available. Accessible to semis. **Rates:** $20 per night, $100 for week w/o feed and hay. **Accommodations:** Quality Inn, Huntingburg (5 miles). Holiday, Comfort, Hampton Inns, Jasper (10 miles).

LA PORTE
Gillerlain Quarter Horse Farm Phone: 219-362-1122
Paul & Joanne Gillerlain
0102 E. 200 North [Severs Road] [46350] **Directions:** 4 miles from Indiana Toll Road; 6 miles from I-94. Call for directions. **Facilities:** Indoor arena, 8 indoor 10' x 10' stalls, turnout with shed, feed/hay available. Call for reservation. **Rates:** $15 per night. **Accommodations:** Ramada, Super 8, Cassidy Motel within 1 mile in LaPorte.

INDIANA

ALL OF OUR STABLES REQUIRE CURRENT NEG. COGGINS, CURRENT HEALTH PAPERS, & OWNERSHIP PAPERS.

MIDDLEBURY
✻ <u>Coneygar</u> Phone: 574-825-5707
Mary Dugdale Hankins
54835 County Road 33 [46540] **Directions:** Located south of Exits 101 and 107 (Middlebury) on Ind. Toll Road 80-90, between US 20 and SR 120. Call for specific directions to stable. **Facilities:** 10 indoor stalls, large dry barnlot and several acres of pasture. Feed/hay and trailer parking available. Boarding only for guests of B&B. **Rates:** $10 for stall per night; $5 for pasture. **Accommodations:** Bed & Breakfast in country home on 40 acres with stable facilities. Includes hearty country breakfast. Near Northern Indiana Amish country.

NASHVILLE
✻ <u>Rawhide Ranch</u> Phone: 812-988-0085
Jenny Peddycord 888-94- Ranch
1292 South State Road 135 S. [47448] **Directions:** 14 miles west of I-65, 13 miles from Columbus, & 1 mile off State Rd. 46. Call for directions. **Facilities:** 32 indoor 14' x 12' stalls, two 1-acre paddocks, 120' x 60' outdoor riding arena, hot walker, riding in nearby Brown County State Park with 100,000 acres of trails. Horses for sale. Hunters & fishermen welcome. **Rates:** $30 per night. **Accommodations:** New cabin on premises available to rent: $100 per night with horse boarded. Also, Brown County Inn, Seasons, & Salt Creek Inn 2 miles away.

NOBLESVILLE
<u>Janet Keesling Stables</u> Phone: 317-773-5482
Janet Keesling
11930 E. 211th [46060] **Directions:** 1/2 mile from US 37; 16 miles from I-465; 7 miles from I-69. Call for directions. **Facilities:** 30 indoor 12' x 12' stalls, 60' x 120' indoor arena, 63' round pen, feed/hay, & trailer parking. Veterinarian within 3 miles. Stallions accepted. Call for reservation. **Rates:** $15 per night. **Accommodations:** Super 8 in Noblesville, 5 miles from stable.

PAOLI
✻ <u>Wilstem Dude Ranch</u> Phone: 812-936-4484
Misty Weisensteiner, Manager or: 812-634-1413
E-mail: wilstem@bluemarble.net Web: www.wilstemguestranch.com
US Hwy 150 West [47454] **Directions:** Call for directions. **Facilities:** 55 indoor 10' x 12' stalls, paddock, large indoor & outdoor arena, 30 miles of trails on a total of 800 acres, feed/hay & trailer parking available. Call for reservation. **Rates:** $15 per night. **Accommodations:** Historic lodge on premises that sleeps 18, plus cabins.

INDIANA   Page 81

*ALL OF OUR STABLES REQUIRE CURRENT NEG. COGGINS,
CURRENT HEALTH PAPERS, & OWNERSHIP PAPERS.*

ROCKPORT
✣ **Ramey Riding Stables** (KY office)   Phone: 812-649-2668
Joan Ramey   Cell: 270-570-3054
Web: www.rameycamps.com   E-mail: jramey66@yahoo.com
2354 S. 200W [47635] Directions: Call for directions. **Facilities:** 32 -8' x 10' indoor stalls, 8 in private, overnight separate barn, 3 acres of pasture turnout, round pen, heated barn, indoor arena, 50' x 100' outdoor arena & miles of riding trails. CHA certified instructor for English, Western & dressage. Residential riding camps, picnic area & electric RV hook-up available. Reservation required. **Rates:** $15 per night. $60 per week. **Accommodations:** Bed & Breakfast on premises with 3 BR condo cottage, finely appointed, sleeps 10, no smoking: $85/night double. Tennis courts, swimming pool, workout rooms, & more in second facility 20 minutes away, Executive Inn, 8 miles away.

TERRE HAUTE
Persimmon Hollow Farm   Phone: 812-299-4754
Bill & Donna Isaacs
2966 E. Harlan Road [47802] Directions: From I-70 and US 41: 7 miles south to Stuckey Pecan Shop; go 9/10 mile east and you will see long white board fence of farm. **Facilities:** 5 indoor 12'x12' stalls, 3 72'x20' holding pens, trailer parking available. Camping facilities 1/2 mile away. Call for reservation. **Rates:** $15 per night. **Accommodations:** Holiday Inn and many others 7 miles from stable.

WEST POINT
Flint Run Stables   Phone: 317-572-2803
Dave & Nancy Shaw
6339 South 700 W [47992] Directions: 2 miles south of SR 25 at West Point on 700W. Call for specific directions. **Facilities:** 8 indoor stalls, 2 - 1/2-acre turnouts, 60' x 85' indoor riding arena, hay & trailer parking available. Electric & water RV hook-ups. Spring water for horses. 10 minutes from Purdue University Large Animal Clinic. Reservations not necessary but call in advance. **Rates:** $15 per night; $75 per week. **Accommodations:** Howard Johnson's & Ramada Inn in Lafayette, 15 minutes away.

- Full Parts & Service Dept.
- Repairs on All Makes of Trailers
- Dealer for Sooner, Titan, Bison, Adam, Bloomer

Parts Shipped Daily

800-776-4635

www.brianstrailersales.com

32 miles west of Indianapolis on US Hwy. 36. Bainbridge IN

Page 82   **IOWA**

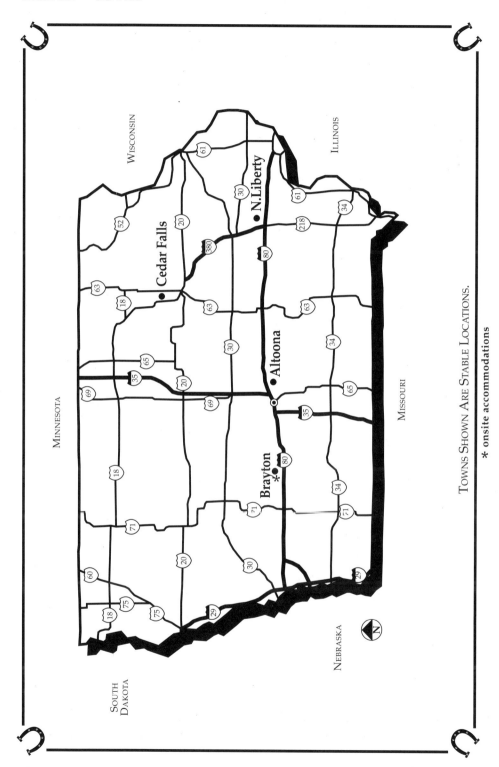

IOWA PAGE 83

ALL OF OUR STABLES REQUIRE CURRENT NEG. COGGINS, CURRENT HEALTH PAPERS, & OWNERSHIP PAPERS.

ALTOONA
Georgia's Arabian Stables   Phone: 515-967-5553
Georgia Campbell
5055 NE 96th Street [50009] **Directions:** 5 miles from I-80: Take Exit 143. At stop sign, turn to go to Altoona then go 1/2 mile to first road on left. Take left and go 4 miles. Turn right at stop sign & go 1/2 mile. Stable is 2nd house on left. **Facilities:** 38 10x12 indoor box stalls, 90' x 100' indoor arena, 3 indoor tie stalls, 3 large turnout pens, 3 round pen, & electric camper hook-up, 20-year 4-H leader, Hay Rack rides, Horse trailer rentals, riding lessons for beginners, advanced and handicapped, Reg. Pinto stud(Black and White), Arabian stud, Stud pen with shed available. Will clean out horse trailer for free. Negative Coggins and current health papers. **Rates:** $15 per night. **Accommodations:** Eight motels within 6 miles of stable. Most will also take pets.

BRAYTON
✱ **Hallock House Bed & Breakfast**   Reservations: 800-945-0663
Guy & Ruth Barton
3265 Jay Avenue [50042] **Directions:** Take Exit 60 off I-80, north to Brayton 3 miles; stable is approx. 1 mile east of Brayton. **Facilities:** 4 inside stalls with outside pens, 2 separate outside pens, 12 inside box stalls available across the road. 1 acre of pasture/turnout, trailer parking available. **Rates:** $10 per night. **Accommodations:** 2 guest rooms on premises. EconoLodge within 4 miles at I-80 and Hwy 71.

CEDAR FALLS
Beahr Ridge Legendary Stables   Phone: 319-988-3021
Richard & Sylvia Beahr
5533 Hudson Road [50613] **Directions:** On Hwy 58 two miles north of Hudson & 3.5 miles off Hwy 20. Call for final directions. **Facilities:** 47 box stalls, 80' x 120' riding arena, 100' outdoor arena, 3 paddocks with covered shelters. Arabian horses for sale. Standing-at-stud: "Rho-sabee." Seven national top-ten titles. Reservations required. **Rates:** $20 per night; $75 per week. **Accommodations:** Motels within 5 miles of stable.

NORTH LIBERTY
Whinnee Acres   Phone: 319-626-6416
Jean Eckhoff, Owner   Fax: 319-626-4717
2237 Scales Bend Road NE [52317] **Directions:** North off Exit 240 of I-80. Call for directions. **Facilities:** 10-12 11'x11' indoor stalls, 63' x 120' indoor arena, 40'x60' pasture/turnout, feed/hay avaliable, trailer parking with electric for $5 per night , trails for riding. **Rates:** $15 per night, **Accommodations:** 7 Motels within 5 miles at Exit 240.

KANSAS

Towns Shown Are Stable Locations.

* onsite accommodations

KANSAS   Page 85

ALL OF OUR STABLES REQUIRE CURRENT NEG. COGGINS, CURRENT HEALTH PAPERS, & OWNERSHIP PAPERS.

AUGUSTA
Unity Morgans Phone: 316-775-1554
Jim & Patricia Michael Fax: 316-775-1554
12698 SW Thunder Road [67010] **Directions:** Three miles southwest of highway 54 & 77. Call for specific directions. **Facilities:** 2 - 12 X 12 indoor stalls. Feed and Hay available. Trailer parking available. Outdoor arena, round pen. **Rates:** $15.00 per night. **Accommodations:** Lehr's Motel and Agusta Inn near by.

White Training Stables Phone: 316-775-3602
Steve White
6404 Southwest Hunter Road [67010] **Directions:** Call for directions. **Facilities:** 20 indoor stalls, indoor arena, feed/hay, & trailer parking available. Specializes in Arabian and Quarter horses. Reservations required. **Rates:** $15 per night; ask for weekly rate. **Accommodations:** Motels 4 miles from stable.

BONNER SPRINGS
Z7 Stables Over Hill Phone: 913-441-8860
Hank & Linda Perrin
3601 S. 142nd Street [66012] **Directions:** Bonner Spring, Hwy 7, exit off of I-70. Call for further directions. **Facilities:** 6 indoor stalls, 3 turnout pens, indoor arena, jumps available. Horseshoeing available and vets nearby. Cannot accommodate tractor-trailers. Minutes away from dog race track & riverboat gambling. **Rates:** $15 per night. **Accommodations:** Motels 15 minutes from stable.

BROOKVILLE
✱ **Castle Rock Ranch** Phone: 785-225-6865
Judy A. Akers
1086 29th Road [67425] **Directions:** Call. **Facilities:** 4 - 12' X 12' indoor stalls. Soft wood chips. Hay available. 2 acre pasture surrounding barn. Trailer parking available. **Rates:** $20 per night. **Accommodations:** Overnight stabling only available to B&B guests. 8 miles from Kanopolis Lake which has 26 miles of riding trails.

COFFEYVILLE
The Horse Center Phone: 620-251-4511
Elton Weeks Home: 251-4649
Rt. 4 [67337] **Directions:** 2 miles north of Hwy 166 West and 1 mile north of Coffeyville Country Club. Call for directions. **Facilities:** 10 indoor 12' x 12' box stalls, 60' x 106' arena, stud pen with run, feed/hay and trailer parking available. Trains driving horses. Reservations required. **Rates:** $6 per night. **Accommodations:** Motels 6.5 miles from stable.

KANSAS

ALL OF OUR STABLES REQUIRE CURRENT NEG. COGGINS, CURRENT HEALTH PAPERS, & OWNERSHIP PAPERS.

EMPORIA
✱ Shamrock Ranch Bed & Board Phone: 620-342-5941
Tim & Carol McLaughlin Web: www.shamrockranch.com
E-mail: shamrock@osprey.net
1323 South Hwy 99 [66801] **Directions:** Call for directions or visit website. **Facilities:** 5 12x12 indoor stalls with straw bedding, hay available, trailer parking with 110/220 plugin available for $10, 100'x100' turnout, large 60'x185' indoor lighted arena/round pen **Rates:** $15 per night. **Accommodations:** Motels within 10 minutes. We provide a bunkhouse that sleeps up to 4 with continental breakfast for $45 per night.

GARDEN CITY
Finney County Fairgrounds Phone: 888-876-3844
Angie Clark   Web: www.finneycounty.org   E-mail: fair@finneycounty.org
209 Lake Ave [67846] **Directions:** Turn west off Hwy 83 South Fairgrounds 1 block west. **Facilities:** 17 indoor stalls, 21 covered stalls, adequate and easy trailer parking. **Rates:** $15 per stall, $10 per electrical hook up, $40 per week, reservations required. **Accommodations:** Best Western, Days Inn, Garden City Inn, Holiday Inn Express, Super 8 within 6 miles.

GREAT BEND
Riverside Stables Phone: 620-793-6523
Dixie Clark E-mail: riversidestablesllc@yahoo.com
222 South Kiowa Road [67530] **Directions:** Located 1/2 mile south of Hwy 56 on east edge of town. Call for directions. **Facilities:** 6 indoor box stalls, 13 stalls with outdoor runs, 75' x 100' indoor lighted arena, 100' x 200' outdoor arena, trailer parking available. Complete tack store on premises. Reservations required. **Rates:** $20 per night. **Accommodations:** Motels 2 miles from stable. Best Western, Super 8, Holiday Inn and others.

SALINA
E BAR Z STABLES Phone: 785-825-7135
Dexter Eggers & Ann Zimmerman Web: www.ebarz.com
E-mail: overnight@ebarz.com
3904 N. Ohio St. [67401] **Directions:** 1.8 miles north of I-70 at exit #253 Ohio Street. Driveway is on the east (right) side of road. Possible road construction in '04. Please call at least 2 hours in advanced if possible. Advanced reservations preferred. **Facilities:** 10 12'x 12' stalls (8 with outdoor runs), 10 double-size outdoor steel pens (40'x 70') with shelter, feed/hay available, trailer parking, circle drive for easy turnaround, 200 acres of riding trails, indoor and out door arenas, foundation quarter horses. **Rates:** $20 per night. Call for weekly rate. **Accommodations:** Several major motels only 3 miles from stable.

KANSAS Page 87

ALL OF OUR STABLES REQUIRE CURRENT NEG. COGGINS, CURRENT HEALTH PAPERS, & OWNERSHIP PAPERS.

TONGANOXIE
Laurel Hill Stables Phone: 913-369-3124
Chris van Anne
19478 Hwy. 16 [66086] **Directions:** 25 miles off of I-70. Call for directions. **Facilities:** 15 indoor stalls, 150' x 200' outdoor arena, indoor arena, separate paddocks & pastures. Feed/hay and parking for big rigs available. Morgan show barn. Trains English & Western & saddleseat lessons. Reservations required. **Rates:** $15 per night; call for weekly rates. **Accommodations:** Motels 3 miles from stable.

TOPEKA
Horse O'Tel, Phone: 785-246-2579
Harold M. Smith
8540 NW Hwy 75 [66618] **Directions:** 8 miles north of I-70 and Hwy 24 on Hwy 75. Milepost #170. **Facilities:** 4 indoor 10' x 10' stalls, no pasture/turnout. Feed/hay and trailer parking available. **Rates:** $10 per night. **Accommodations:** Motel 6 & Holiday Inn in Topeka, 8 miles from stable.

WAKEENEY
✣ **Saline River Hunting Lodge & Guide Services, Inc.** Phone: 785-743-5878
Kevin & Julia Struss E-mail: knss@ruraltel.net
R.R. 2 Box 14B [67672] Web: www.salinelodge.com
Directions: I-70, exit 128, 7 miles north on Hwy 283, 2 miles west on county road 422. **Facilities:** Indoor pens for up to 3 horses, 2 large and 2 small outdoor holding pens. Feed/hay available, trailer and camper parking with electrical and water hookups, Dog kennels also available. 45 acre native prairie. **Rates:** $10/horse/night. **Accommodations:** Very nice lodge for up to 6 individuals, full home cooked breakfast included with lodging. Reservations required.

✣ **Thistle Hill Bed & Breakfast** Phone: 785-743-2644
Dave & Mary Hendricks Web: www.thistlehillonline.com
Rt. 1, Box 93 [67672] **Directions:** 1.5 miles from I-70 halfway between Kansas City and Denver. Complete directions given at time of reservation. **Facilities:** 2 indoor pens, 4 large outdoor holding pens. No feed/hay. Trailer parking available. Pets subject to restraint. Accommodations not suitable for pets indoors. Reservations required. Smoking restricted. **Rates:** $15 per night per horse. **Accommodations:** Bed & Breakfast is a comfortable, secluded cedar farm home situated on 320 acres. Wonderful, homemade complimentary breakfast included. $59-$75 per room/double occupancy; $59 single occupancy. Deposit requested to guarantee reservation.

Page 88 KENTUCKY

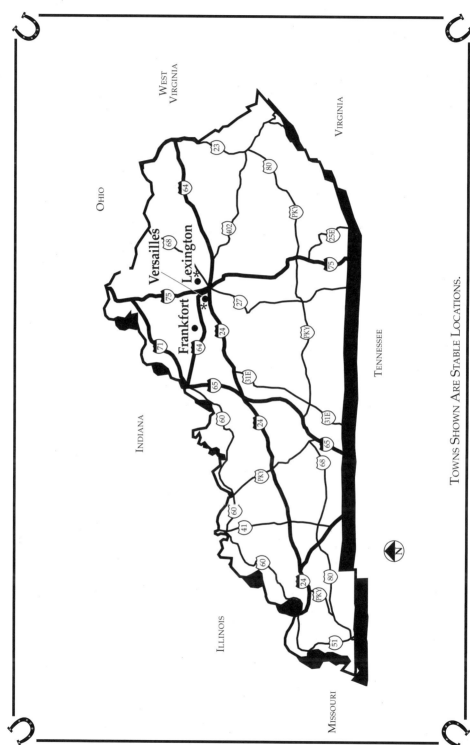

KENTUCKY Page 89

<u>*ALL OF OUR STABLES REQUIRE CURRENT NEG. COGGINS, CURRENT HEALTH PAPERS, & OWNERSHIP PAPERS.*</u>

FRANKFORT
<u>Jackson Stables</u>　　　　　　　　　　　　　Phone: 502-223-5412
William & John Jackson　　　　or: Wm: 502-695-2853; John: 502-695-1298
126 Shadrick Ferry Road [40601] **Directions:** 10 miles off of I-64. Call for directions. **Facilities:** 8 indoor stalls, 3-acre pasture, 3-acre turnout area, 86' x 210' outdoor riding ring, indoor riding ring, feed/hay & trailer parking available. Horse training, buying & selling. Specializing in Tennessee Walkers. Call for reservation. **Rates:** $15 per night. **Accommodations:** Holiday Inn 5 miles.

FRANKFORT/VERSAILLES
<u>Lakeside Arena</u>　　　　　　　　　　　　　Phone: 859-489-4885
Bruce Brown　　　　　　　　　　　　Web: www.LakesideArena.com
1385 Duncan Millville Rd [40601] **Directions:** I-64 Exit #58, South on US 60 1/4 mile to S.R. 1681 West. Drive 1.2 miles, farm on left. **Facilities:** 200- 10x10 indoor stalls. Feed/hay, trailer parking and pasture available. Reservations only. **Rates:** $20 per night. **Accommodations:** 60 RV hook-ups available. 1-1/2 mile to Best Western Parkside 502-695-6111.

LEXINGTON
✻ <u>Spring Bay Farm Bed & Stall</u>　　　　　　　　Phone: 859-231-8702
John S. & M. Stanley Wiggs　　　　　　E-mail: J.wiggs1@insightbb.com
1249 Greendale Road [40511] **Directions:** I-64/75, Exit 115-Newton pike Hwy 922 to Circle #4, to exit 7. Leestown Rd Hwy 421, turn right, 2nd light turn right onto Greendale Rd. 1.7 miles on left. **Facilities:** 3 stalls in separate guest barn 10 X 12 box stalls in wood barn with clay floors, bedding with shavings. Trailer parking available. 1-4 acre pasture, 1 holding paddock. Call for reservations for stalls before 8pm, for reservations for room 2 weeks ahead, recorder on all times. Within 5 miles to two major horse parks. Non-refundable deposits required for rooms and /or multiple stalls. 1880's restored Victorian Farm House, call for more details. No pets or smoking allowed in house. **Rates:** $25 per night per horse. **Accommodations:** La Guinta, Holiday Inn North, Four Points Sheraton, Embassy Suites, Marriott Resort all within 5 miles.

VERSAILLES
✻ <u>River Mountain Farm B&B</u>　　　　　　　　Office: 859-873-1604
Elaine & Trey Schott, DVM　　　　　　　　Fax: 859-879-3337
E-mail: RMFhorses@msn.com
3085 Troy Park [40383] **Directions:** I-64 or I-75 (Newtown Park Route 922 exit). Follow signs to BG Parkway. BG Parkway to first exit, Versailles. Turn left, go 1 mile, farm on right. Great access to Kentucky Horse Park **Facilities:** 45 - 12' X 12' wood stalls. Feed, hay and trailer parking available. Total 100 acres with 15 paddocks. English/Polo riding lessons; University of Kentucky Riding Team Training Center; Hunter Jumper our specialty. Private turnout available. **Rates:** $20 per night. $125 per week. **Accommodations:** Best Western, US 60, Frankfort. Comfort Inn, Harrodsburg Rd, Lexington. Bed & Breakfast available on farm: $85 per night (sleeps 4, full kitchen, private entrance).

LOUISIANA

Towns Shown Are Stable Locations.

* onsite accommodations

LOUISIANA Page 91

ALL OF OUR STABLES REQUIRE CURRENT NEG. COGGINS, CURRENT HEALTH PAPERS, & OWNERSHIP PAPERS.

BATON ROUGE
Farr Park Horse Activity Center Phone: 225-769-7805
Karen Toney, Equestrian Manager
6402 River Road [70820] **Directions:** 2.5 miles south of LSU campus on River Road; 6 miles south of downtown Baton Rouge; 5.5 miles south of I-10 College Drive Exit. Call for specific directions. **Facilities:** 230 outdoor covered stalls, 121' x 244' indoor arena with grandstand seating for 1,500, wash racks, 140' x 300' outdoor arena, polo field, 24-hour staff presence, 151-unit RV area with electric, water hookups for $10 per night, plus bathrooms, showers, & laundry. Center is site of rodeos, team-penning, dressage & jumping events. Boarding arrangements must be made 24 hours in advance. **Rates:** RV $12 per night, stalls $15 per night **Accommodations:** Sheraton Baton Rouge (4 miles), Comfort Inn (7 miles).

CARENCRO
Traders Rest Farm Phone: 337-234-2382
Don Stemmans Fax: 337-234-2383
E-mail: info@stemmans.com Web: www.stemmans.com
P.O. Box 156 [70520] **Directions:** 2 miles from I-49 near Lafayette. 1 mile north of I-10 at Ambassador Caffrey, go north, at T - take left, road will curve once, at 2nd T - turn right. After 3rd curve, our farm is on the right. **Facilities:** 130 indoor stalls, pasture/turnout available, feed/hay, trailer parking on premises. Thoroughbred & quarter horse stallions, racing facility. **Rates:** $15 per night. **Accommodations:** Many hotels and restaurants nearby.

HAUGHTON
Double Rainbow Equestrian Center Phone: 318-949-9133
Sig North or: 318-949-9948
E-mail: dblrainbowf@tol.com Web: www.DoubleRainbowFarm.com
1860 Adner Road [71037] **Directions:** 3 miles from I-20 and 220 Loop. Also 3 miles from Louisiana Downs. Call for directions. **Facilities:** 150' x 288' lighted covered arena, 50 indoor stalls in 6 barns on 146-acre Center, numerous 1/2 to 30-acre paddocks, 10 wash racks with H/C water, racetrack, electric walking wheel at each barn, 2 outdoor arenas, 2 round pens, feed/hay & buckets available, trailer parking on site. 24-hr security. Farrier on premises. Multi-disciplinary riding & training programs offered at Center including therapeutic riding for the handicapped. Call for reservation. **Rates:** $15 per night; $75 per week. **Accommodations:** leBossier Motel & Grande Isle Motel both 4 miles from Center.

SHREVEPORT
State Fair of Louisiana Phone: 318-635-1361
3701 Hudson Street [71109] **Directions:** Call for directions. Off I-20. **Facilities:** 132 permanent stalls, 40 in cattle barn, trailer parking available ($15.00 per night), feed store close by. Reservations required; gates locked at 5 P.M. Fair dates October 23 -- November 8; very full at that time. **Rates:** $15 per stall per night. **Accommodations:** Cluster of motels near airport (1 mile).

LOUISIANA

ALL OF OUR STABLES REQUIRE CURRENT NEG. COGGINS, CURRENT HEALTH PAPERS, & OWNERSHIP PAPERS.

SLIDELL
Lewis Stables　　　　　　　　　　　　　　Phone: 985-643-8025
Bob Lewis　　　　　　　E-mail: lewistrl@cmg.net　　Fax: 985-643-6791
80 Tortoise Street [70461]   **Directions:** From I-10: Take Exit 266; go east 1 mile to first traffic light; take left (north) & go to 3rd stop sign; take right & go 2 blocks to stable. **Facilities:** 130 indoor stalls, 6-8 holding pens, Purina dealer on site. Can handle large transports. Trailer sales, service, & rentals on site as well as a tack store. Horse transportation business from this location primarily serving the Southeast but will transport nationwide on request. Personal & speedy service their specialty. Advance reservations required. **Rates:** $15 per night; weekly rate negotiable. **Accommodations:** Ramada Inn, LaQuinta, within 1.5 miles.

NOTES AND REMINDERS

Page 94 MAINE

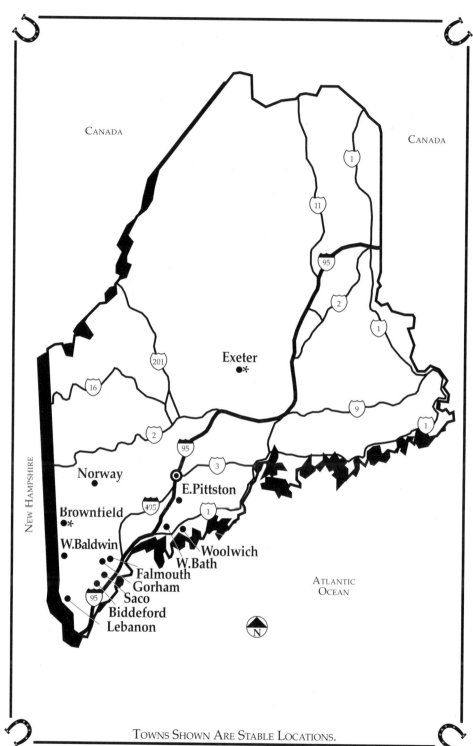

Towns Shown Are Stable Locations.
* onsite accommodations

MAINE PAGE 95

ALL OF OUR STABLES REQUIRE CURRENT NEG. COGGINS, CURRENT HEALTH PAPERS, & OWNERSHIP PAPERS.

BIDDEFORD
Bush Brook Farm   Phone: 207-284-7721
Mona Jerome   Home: 207-284-8311
4463 West Street [04005] **Directions:** Exit 4 (Biddeford) off of I-95. Call for directions. **Facilities:** 4 indoor stalls, feed/hay, trailer parking available. Jumps & trails for riding. Near 3 beaches. **Rates:** $20 per night. Call for weekly rate. **Accommodations:** RV park 1/2 mile and motels 3 miles from stable.

BROWNFIELD
✶ The Foothills Farm B & B   Phone: 207-935-3799
Kevin Early & Theresa Rovere
RR 1, Box 598 [04010] **Directions:** 9 miles from Route 302 & 10 miles from Route 16. Call for directions. **Facilities:** 3 indoor, 2 outdoor lean-to stalls, pasture, and riding trails available. No smoking preferred. Reservations required. 10 miles from Conway, NH and 9 miles from Fryeburg. Must stay at B&B. **Rates:** $15 per night. **Accommodations:** B&B: $42-$48 double occupancy with shared bath.

E. PITTSTON
Woodrose Farm   Phone: 207-582-6315
Karen Lyons
Mast Road [04345] **Directions:** Rt. 1 to Rt. 27. 5 miles off Rt. 27. Call for further directions. **Facilities:** 3 indoor stalls, indoor arena, pasture, paddocks, & pens. Dog boarding OK. Please call ahead. Dressage lessons available. **Rates:** $10 per night. **Accommodations:** Motels 10 miles from stable.

EXETER
✶ Sirsarg Stables   Phone: 207-379-2776
Noel Sirabella
Champion Road [04435] **Directions:** Newport Exit off I-95 to Rte 11; approx. 13 miles off exit. **Facilities:** 5 indoor 10' x 10' box stalls, 3 indoor 5' x 8' walk-in stalls, 16' x 40' lean-to, 50' x 70' paddock, round pen, 150' x 100' outdoor riding ring, barrel course, 14- & 10-acre fenced pastures, feed/hay, trailer parking available. Trail riding available. Farrier, grain and feed store, vet in town. Buys and sells horses, generally quarter horses. **Rates:** $12 per night; weekly rate available. **Accommodations:** Spare bedrooms on premises; motels in Newport (13 miles away) and Bangor (18 miles away).

FALMOUTH
Norton Farms   Phone: 207-797-7577
Lori Graffam
613 Blackstrap Road [04105] **Directions:** Exit 10 off of Maine Turnpike (I-95). Call for directions. **Facilities:** 28 indoor stalls, 4 paddocks, 4 pastures, walker. Horse trailer parking available. Breeds & trains Standardbreds. Please call ahead for reservations. **Rates:** $15 per night. **Accommodations:** Motel 8 miles from stable.

MAINE

ALL OF OUR STABLES REQUIRE CURRENT NEG. COGGINS, CURRENT HEALTH PAPERS, & OWNERSHIP PAPERS.

GORHAM
Kents' Stables   Phone: 207-839-5351
Lisa Kents   207-839-6428
726 Fort Hill Road [04038]
Directions: Exit 8 off of I-95. Call for directions. **Facilities:** 5 indoor stalls, paddock area, indoor & outdoor arenas, outside stadium arena, outside dressage arena, cross-country course, miles of trail riding including beach riding. Lessons available. Call for reservations. **Rates:** $25 per night. **Accommodations:** Motels within 5 miles of stable.

LEBANON
Menomin Meadow   Phone: 207-457-1774
Corine Crossmon
41 Columbus Circle [04027] **Directions:** Call for directions. **Facilities:** Up to 10 indoor stalls in a new barn, some varied size paddocks, feed/hay at additional cost. Vaccinations for E&WEE, tetanus, flu, rhino, & Potomac in last 3 months. Payment upon arrival by cash or travelers check only. **Rates:** $15 for 10' x 12' stall; larger stalls available; call for weekly rate. **Accommodations:** Cardinal Ranch Motel in Rochester, NH, about 7 miles from stable.

NORWAY
Hidden Brook Farm   Phone: 207-743-6546
Beth & Paul Brainerd
RFD 1, Box 968 Howe Road [04268] **Directions:** 8 miles from Rt. 26 & 30 minutes from Fryeburg. Call for directions. **Facilities:** 20 indoor 12' x 12' stalls, 20m x 60m wood-fenced grass pasture areas. Farrier on premises. Dressage & combined training. Call for reservations. **Rates:** $15 per night; $75 weekly rate. **Accommodations:** Inn Town Motel in Norway, 6 miles away.

SACO
Breezy Meadow Horse Farm   Phone: 207-284-9409
Doreen Metcalf   207-284-4074
184 Buxton Road [04072] **Directions:** 2 miles off of I-95. Call for directions. **Facilities:** 6 indoor stalls, indoor ring, 2 outdoor rings, 4 small paddocks, 5- & 10-acre paddocks. Horse training & sales. Call for reservations. **Rates:** $15 per night. **Accommodations:** Motel 2 miles from stable.

WEST BATH
Whorff Stables   Phone: 207-443-3965
Rhonda Whorff
Foster's Point Road [04530] **Directions:** Call for directions. **Facilities:** 6 indoor stalls, 80' x 100' riding ring, large pasture, feed/hay available. Trail rides and lessons offered at stable. Call ahead for reservations. **Rates:** $20 per night. **Accommodations:** Motels 3 miles from stable.

MAINE

ALL OF OUR STABLES REQUIRE CURRENT NEG. COGGINS, CURRENT HEALTH PAPERS, & OWNERSHIP PAPERS.

WOOLWICH
Gallant Morgan Horse Farm **Phone:** 207-443-4170
Margarite Gallant
45 Temple Road [04579] **Directions:** Call for directions. **Facilities:** 4 indoor stalls, indoor & outdoor rings, farrier available. Tack shop on premises. Morgans for sale. Farm offers a summer horse camp for girls. Please call for reservations. **Rates:** $15 per night. **Accommodations:** Motels 6 miles from stable.

MARYLAND

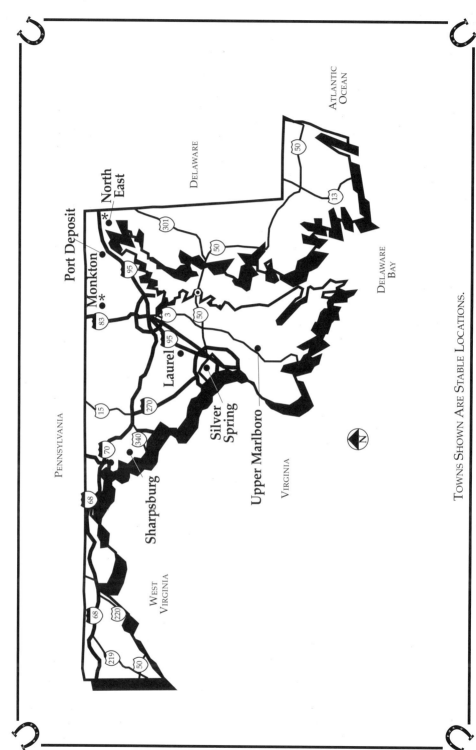

Towns Shown Are Stable Locations.

* onsite accommodations

MARYLAND Page 99

ALL OF OUR STABLES REQUIRE CURRENT NEG. COGGINS, CURRENT HEALTH PAPERS, & OWNERSHIP PAPERS.

LAUREL
Columbia Horse Center   Phone: 301-776-5850
Nanci Steveson
10400 Gorman Road [20723] **Directions:** Exit 35B off of I-95. Call for directions. **Facilities:** 20 indoor stalls, 2 large outdoor rings, 2 indoor arenas, 12 paddocks on 88 acres. Lessons & training for huntseat only. Call for reservation. **Rates:** $25 per night. **Accommodations:** Motels 5 miles from stable.

MONKTON
✱ **Upper Crondall Farm**   Phone: 410-472-4528
Jack & Betsy Ensor
16909 Gerting Road [21111] **Directions:** I-83 to Exit 27, east 1 mile to stop light, right onto York Road, 1 block to left onto Monkton Road, go 3 miles to Sheppard Road, follow Sheppard Road for 2 miles, left at first stop sign (Gerting Road), farm is first house on right. **Facilities:** 12' x 12' indoor stall, pasture/turnout, hay & trailer parking available. Call for reservation. **Rates:** $15 per night. **Accommodations:** Bed & breakfast on premises, $85 per night, fireplace, private bath, air conditioning.

NORTH EAST
✱ **Tailwinds Farm**   Phone: 410-658-8187
Ted & JoAnn Dawson   Web: fairwindsstables.com
E-mail: jdawson@fairwindsstables.com
41 Tailwinds Lane [21901] **Directions:** From I-95 take exit 100 (MD), this will be Rt. 272. Take 272 north 3.5 miles, farm is on the right. **Facilities:** 25 12'x12' indoor stalls, 25 acres of separated pastures, feed/hay available, ample trailer parking. Indoor and outdoor rings, trail rides, riding lessons and carriage rides available. Farm is located 10 minutes from Fair Hill State Park, featuring 5,000 acres with riding trails. **Rates:** $25 per night box stall. $15 per night turnout only. **Accommodations:** Tailwinds is a Bed & Breakfast, 2 rooms available at $75 per night. Crystal Inn in North East just 3 miles away.

PORT DEPOSIT
Anchor and Hope Farm   Phone: 410-378-4081
Edwin Merryman
P.O. Box 342 [21904] **Directions:** Located 4 miles off of I-95: Please call ahead for directions. **Facilities:** 3 indoor stalls, paddocks, pasture/turnout, feed/hay & trailer parking available. **Rates:** $30 per night. **Accommodations:** Comfort Inn at I-95, 4 miles from farm.

MARYLAND

ALL OF OUR STABLES REQUIRE CURRENT NEG. COGGINS, CURRENT HEALTH PAPERS, & OWNERSHIP PAPERS.

SHARPSBURG
Poor Boy Stables Phone: 301-223-9089
Raymond Ramsey
16419 Woburn Road [21782]   **Directions:** 5 miles off I-81 & I-70. Call for directions. **Facilities:** 22 indoor box stalls; 6-, 2-, & 1-acre pasture/turnout areas; 85' x 165' indoor arena; 6 outside paddocks; rings; and riding trails. Large rigs OK. Lessons & carriage rides, sleigh & hay rides, etc. Also buying & selling of horses, specializing in child-proof horses. Located 4 miles from Antietam Battlefield. Call for reservation. **Rates:** $15 per night. **Accommodations:** Days Inn in Williamsport, 6 miles from stable.

SILVER SPRING
Woodland Horse Center Phone: 301-421-9156
Michael Smith
16301 New Hampshire Avenue [20905]   **Directions:** 8 miles from Beltway. Call for directions. **Facilities:** 5 indoor stalls, pasture & small turnout paddock, 5 - 2- to 5-acre pastures. No stallions. OVERNIGHT BOARDING IN AN EMERGENCY SITUATION ONLY. **Rates:** $15 per night. **Accommodations:** Many motels 15 minutes away in Silver Spring.

UPPER MARLBORO
Prince George's Equestrian Center Phone: 301-952-7900
Maryland National Capital Park and
Planning Commission
Liz Yewell
14900 Pennsylvania Avenue (MD Rte. 4) [20772]   **Directions:** Call for directions. **Facilities:** 240 - 12' x 12' block barns. Stabling available by appointment only. Availability limited on weekends between mid March & November due to horse shows. Ship-ins to reserve stalls in advance, feed & bedding can be delivered if arranged in advance. Trailer parking available. Negative coggins required. **Rates:** $15 per night, plus $6 cleaning if shavings are used. **Accommodations:** Call for information. Hampton Inn/Largo.

NOTES AND REMINDERS

MASSACHUSETTS

Towns Shown Are Stable Locations.
* onsite accommodations

MASSACHUSETTS PAGE 103

ALL OF OUR STABLES REQUIRE CURRENT NEG. COGGINS, CURRENT HEALTH PAPERS, & OWNERSHIP PAPERS.

CANTON
Canton Equestrian Center Phone: 617-821-5527
Mary Hughes 617-828-0335
1095 Randolph Street [02021] **Directions:** Take Rt. 138 off of Rt. 128 (I-95). Call for further directions. **Facilities:** 35 indoor stalls, indoor arena, outdoor arena, jump course, 3 fenced turnouts, miles of riding trails nearby. English & Western, & hunter/jumper lessons at all levels. Call for reservations. **Rates:** $15 per night. **Accommodations:** Holiday Inn 2 miles from stable.

CARVER
Celtic Wind Stables Phone: 508-728-8675
Tara Pratt
58 Plymouth St [02360] **Directions:** Call for directions. **Facilities:** 10 stalls, riding arena, turn out, trail riding, training and lessons. Feed and hay available **Rates:** $15 per night **Accommodations:** Many motels in historic Plymouth area.

GREENFIELD
Meadowcrest Farm Phone: 413-773-7842
Jim & Rita Adams
290 Leyden Road [01301] **Directions:** 2 miles off of I-91 at Exit 26. Call for directions. **Facilities:** 12 indoor 12' x 12' box stalls, outdoor lighted ring, pasture/turnout on 64 acres. Meadowcrest is also a Christmas tree farm with hay rides & sleigh rides drawn by a team of Percherons. Many activities offered at this scenic location. Call for reservation. **Rates:** $20 per night. **Accommodations:** Howard Johnson's 2 miles from stable at I-91.

LUNENBURG
Pine Fall Farm Phone: 508-582-7748
Tammy McAlpine Evenings: 508-582-7601
271 Elmwood Road [01462] **Directions:** Located off of Rt. 2A. Call for directions. **Facilities:** 24 indoor stalls, 60' x 120' indoor arena, fenced grass paddocks, 70' x 140' outdoor arena. English & Western training for horses & riders. Also horse sales. Call for reservation. **Rates:** $20 per night. **Accommodations:** Motels 15 miles away.

NEW BRAINTREE
* **Ash Lane Farm** Phone: 508-867-9927
Mary Kay Newton
Havens Road [01531-0192] **Directions:** Sturbridge Exit off I-90, Rte. 20 east, left onto Rte. 49, left onto Rte. 9, right onto Rte. 67, left at Reeds Country Store. Second farm on right. **Facilities:** 4 indoor 12' x 12' stalls, 2- and 3-acre fields, indoor arena, outdoor sand ring, feed/hay, trailer parking available. Arabian breeding farm, foals for sale. Arabian stallion standing-at-stud: "Ganesh." **Rates:** $25 per night. **Accommodations:** Bed & breakfast on premises, $40 per person, $75 per couple. Motels available near Sturbridge Village.

OUR HANDS FOR HORSES

Equine Massage & Specialized Grooming
Butch & Jean Davidson
Acushnet, MA
508-990-0817

THE BENEFITS OF EQUINE MASSAGE

Equine Massage is therapy for horses. It is commonly used to enhance performance levels and endurance, to prevent injury, and to aid in the rate of recovery. The benefits of Equine massage can be tremendous. Over time the horses physical, mental and emotional properties can be positively enhanced.

The type of massage we utilize is similar to Shiatzu. This type of massage works on the energy lines or meridians of the body, which all beings possess. We also incorporate certain aspects of Swedish and Sports massage, where it best suits the horse during the treatment. The massage is never forced upon the horse, and can cause no harm. Due to the different uses and lifestyles of each horse, great understanding and patience is shown during each session. We humans affect our horses in both positive and negative ways in our daily handling of them. Every riding discipline takes its toll on the physical body and the emotional and mental status of our horses. Horses are great athletes and truly wish to succeed and please us through what we ask them to do. It is our responsibility to provide care for them that will enable this to be achieved.

We are dedicated to the well being of all horses. Our desire is to help provide them with the soundest bodies, minds, and emotional well being each can posess. Our very effective massage and specialized grooming can help your equine achieve this to the best of their ability, regardless of your riding discipline. Treat them like the athletes they are. Give them the advantage and the advantage will be yours. For the love and performance of your great athletes, call us.

The stress, both physical and mental, caused by trailering your horses, can be great. We can alleviate this stress. Call us, your horses will thank you. Advance calls preferred.

Butch Davidson is a 1994 graduate of the Bancroft School of Massage Therapy in Worcester, Massachusetts. He is a 2003 graduate of Geary Whittings School of Equine Sports Massage Therapy in Douglas City, California. He has experience with a variety of breeds, of varying ages.

Jean Berberian-Davidson, wife and partner, has experience with horses ranging over almost thirty years. She is trusted explicitly with the handling of the horse being massaged. She is a master of grooming, for the horses health and well being.

We make a great team.

We are currently working on horses at Perry Paquette Farm located in East Fairhaven, Massachusetts. Home of Champion Horses and Champion Riders. It is owned and operated by Reggie and Dottie Paquette, she can be reached at **508-993-7578**.

MASSACHUSETTS Page 105

ALL OF OUR STABLES REQUIRE CURRENT NEG. COGGINS, CURRENT HEALTH PAPERS, & OWNERSHIP PAPERS.

NORTH ANDOVER
Andover Riding Academy Phone: 508-683-6552
Frank Fiore Evenings: 508-683-9387
16 Berry Street, Rt. 114 [01845] **Directions:** Centrally located near I-93, I-495, & I-95. Call for directions. **Facilities:** 80 indoor stalls at a complete horse facility. Feed/hay & trailer parking available. Call for reservation. **Rates:** $20 per night. **Accommodations:** Candlelight Motel within 1 mile of stable.

PEPPERELL
Twin Pine Farm Phone: 508-433-5252
Cathy & Toby Tyler
34 Jewett Street [01463] **Directions:** 1 mile from Rt. 113 & Rt. 111. Call for directions. **Facilities:** 25 indoor stalls, 6 outside paddocks, 70' x 200' indoor arena, 150' x 225' outdoor arena, wash racks, riding trails, jumping arena. Training for horses & riders in English, Western, jumping, dressage, & Western reining. Overnight boarding in emergencies only. **Rates:** $25 per night. **Accommodations:** Motels 8 miles from stable.

PROVINCETOWN
Bayberry Hollow Farm Phone: 508-487-5600
Chris Lorenz
P.O. Box 1427 [02657] **Directions:** Rt. 6A east and take 3rd Provincetown exit on Shankpanter Road; take right onto Bradford St.; go 1/2 mile & take right onto West Vine St. extension. See Farm sign. **Facilities:** 6 indoor straight stalls, 4 paddocks, trailer parking available. Massage therapy available. Hay provided. Pony rides. Call for reservation. **Rates:** $15 per night; $75 per week. **Accommodations:** Motels nearby.

REHOBOTH
Gilbert's Tree Farm Bed & Breakfast Phone: 508-252-6416
✱ Jeanne D. Gilbert Web: www.gilbertsbb.com
30 Spring Street [02769] **Directions:** Call for directions. Located 3.5 miles from Rte. 195. **Facilities:** 5 indoor box stalls, 50' x 50' pasture, feed/hay, trailer parking available. Riding trails through 100 acres. Jeanne is licensed riding instructor. **Rates:** $25 one night, $20 two or more consecutive nights, weekly $125. **Accommodations:** Non-smoking bed & breakfast in 150-year-old home with in-ground pool and full breakfast. Also a secluded cabin on premises.

REVERE
Revere-Saugus Riding Academy Phone: 617-322-7788
Mary Ward 617-324-1594
122 Morris Street [02151] **Directions:** Located right off of Rt. 1. Call for directions. **Facilities:** 15 indoor box stalls, 4 turnout paddocks, large turnout area, outside hunt course, riding trails, feed/hay & trailer parking. Training & riding lessons and sales of horses. Call for reservation. **Rates:** $15 per night; $100 per week with group discount. **Accommodations:** Many motels on Rt. 1 within 1 mile.

MASSACHUSETTS

ALL OF OUR STABLES REQUIRE CURRENT NEG. COGGINS, CURRENT HEALTH PAPERS, & OWNERSHIP PAPERS.

ROCHESTER
Bowen Lane Stable　　　　　　　　　　　　Phone: 508-763-1741
Joseph, Diane & Curry DeLowery
68 Bowen's Lane [02770] **Directions:** Rochester exit off 195 onto Route 105. Bowen Lane on left 1/2 miles from Rochester Town Hall/Plum Corner. **Facilities:** 20 full size indoor stalls. Feed & hay available. Trailer parking available. 2 - 600' paddocks; smaller areas will be available. Indoor arena, outdoor dressage area. Acres of trail riding available plus lots of woodlands and corn fields. **Rates:** $16.95 per night. **Accommodations:** 5 miles away in Marion, Ma.

RUTLAND
Holiday Acres Equestrian Center　　　　　Phone: 508-886-6896
Deborah & Clifton Hunt
331 Main Street [01543] **Directions:** Located on Rt. 122A. Call for directions. **Facilities:** 6 indoor stalls, nine - 1/8-acre paddocks, boarded riding ring, indoor riding arena. No stallions. Hunter shows held at Center. Call ahead to check availability. **Rates:** $15 per night; $75 per week. **Accommodations:** Motels in Worcester, 20 minutes away.

WEST NEWBURY
Rowland Meadow Farm　　　　　　　　　　Phone: 978-363-8128
Jefferson &Lauren Rowland
69 Crane Neck Street [01985] **Directions:** Rt 95 to exit 57, West on rt 115 5.2 miles to Crane Neck Street. **Facilities:** 7 indoor stalls, 5 outdoor grassy paddocks, feeed/hay available, small trailer parking, No Stallions . **Rates:** Call for rates. **Accommodations:** Hampton Inn-Hampton, NH (10 miles), Best Western-Haverhill, MA (5 miles)

WESTFIELD
AJ Stables　　　　　　　　　　　　　　　　Phone: 413-562-5974
Tammy & Irene Lowe
1040 E. Mountain Road [01086] **Directions:** Exit 3 off of Mass. Turnpike (I-90). Also easy access off of I-91. Call for directions. **Facilities:** 10 indoor box stalls, large rodeo riding arena, pasture, paddocks. Blacksmith on premises. Training and breaking of horses at all levels. Trains racehorses. Also, horse transportation available serving New England and New York. Call for information and reservations. **Rates:** $25 per night. **Accommodations:** Motels 2 miles from stable.

NOTES AND REMINDERS

Page 108 MICHIGAN

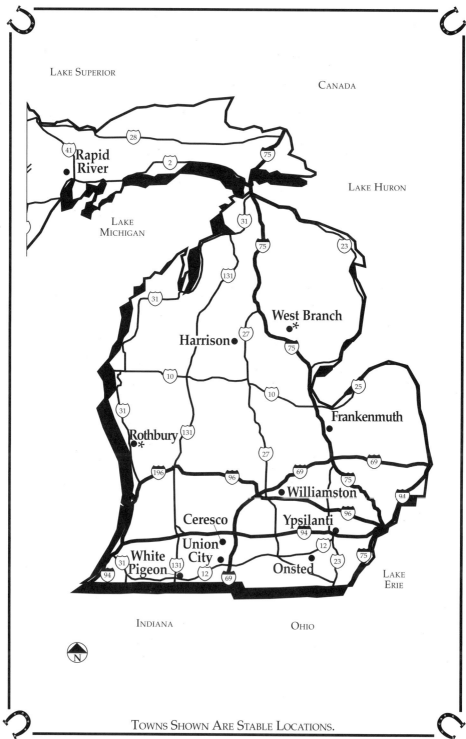

Towns Shown Are Stable Locations.

* onsite accommodations

MICHIGAN Page 109

ALL OF OUR STABLES REQUIRE CURRENT NEG. COGGINS, CURRENT HEALTH PAPERS, & OWNERSHIP PAPERS.

CERESCO
Harper Creek Stables   Phone: 269-979-1554
Deenna Hamilton
11432 8 1/2 Mile Road [49033] **Directions:** Exit 100 off of I-94. Call for directions. **Facilities:** 45 indoor box stalls, 3 outdoor paddocks, 60' x 120' indoor arena, 12 acres of fenced pasture at this boarding facility. Call for reservation. **Rates:** $20 per night. **Accommodations:** Comfort Inn 4 miles.

FRANKENMUTH
Marigold Stable   Phone: 989-652-8761
Teri Cox
7670 E. Curtis Road [48734] **Directions:** Frankenmuth/Birch Run Exit off of I-75. About 7.5 miles on Dixie Hwy. Call for directions. **Facilities:** 65 indoor & outdoor stalls, paddock areas, all levels of lessons in all seats. Feed/hay at extra charge, trailer parking in summer but limited in winter. Reservations required. **Rates:** $10 per night. **Accommodations:** Many motels within 5 miles of stable.

HARRISON
Horse'n Around Tack Shop & Equestrian Center   Phone: 231-539-8500
Joyce Hamsher
8400 N. Bass Lake Road [48625] **Directions:** Go from US 27 to Old 27: 6 miles north of Harrison on Old 27; left on Long Lake Road; go 1 mile to Bass Lake; turn right. Stable is on right side of road. **Facilities:** 22 indoor stalls, 7 pastures & lots, 60' x 120' indoor arena, 110' x 200' outdoor arena. Tack shop & feed store on premises, trailer parking. Mobile tack shop goes to rodeos, shows, etc. Western & English training for horses & riders available. **Rates:** $15 for indoor stall, $10 for outdoor per night. **Accommodations:** Deer Trail Motel 6 miles from Center & rustic camping on site if self-contained.

ONSTED
Green Hills Stables   Phone: 517-467-7614
J. Scott Schultz
10500 Stephenson Road [49265] **Directions:** Located within 5 miles of US 12, US 223, & M-50. Call for further directions. **Facilities:** 6 indoor 10' x12' stalls, 2 outbarns with paddocks, 4 pasture/turnout areas, 150' x 300' lighted arena used for roping, nature trails for riding, 12% sweet feed plus hay available and lots of trailer parking. **Rates:** $20 per night. Call for weekly rate. **Accommodations:** Located in the heart of the Irish hills with many motels and things to do. Call for more information.

RAPID RIVER
Spruce Winds Farm   Phone: 906-474-9701
Deb & Craig Olsen, Pat & GaryDeGrave   906-474-6520
9811 Y.25 Lane [49878] **Directions:** Call for Reservations and Directions **Facilities:** 6 tie stalls, 3 indoor 10x12 Box stalls. Feed/hay, trailer parking available. 1.5 acre turnout, 1 150'x150' pen. Adjacent to Hiawatha Forest and numerous trails. **Rates:** $10 per night. **Accommodations:** Several small motels within 6 miles.

Page 110 MICHIGAN

ALL OF OUR STABLES REQUIRE CURRENT NEG. COGGINS, CURRENT HEALTH PAPERS, & OWNERSHIP PAPERS.

ROTHBURY
✻ **Double JJ Resort Ranch**   Phone: 231-894-4444
Bob Lipsitz   800-DoubleJ
5900 Water Road [49452] **Directions:** Located 20 miles north of Muskegon. Exit off U.S. 31 at the Winston Road, Rothbury Exit. Turn east on Winston Road for .5 miles, north on Water Road. Ranch office is one mile on right. **Facilities:** 10 indoor 12' x 12' stalls, many 12' x 30' turnout pastures on 1000 acres of riding trails, feed/hay, trailer parking available. Dude ranch, full-service resort, championship golf course, weekly rodeo, steak & breakfast rides. **Rates:** $10 per night; $25 minimum. **Accommodations:** RV Park, Hotel, Cabins & Condominiums on premises. Call for more information.

UNION
Camp Bellowood Recreation World   Phone: 269-641-7792
Dan Galbraith
14260 East US 12 [49130] **Directions:** Located off of Indiana Toll Road (I-80/90) at Elkhart or Bristol exits. Call for directions. **Facilities:** 20 indoor stalls, indoor and outdoor arenas, 300 acres of riding trails, feed/hay & trailer parking available. Campground facilities. 24-48 hours notice required. **Rates:** Negotiable. **Accommodations:** Several motels on US 12, eight miles away.

WHITE PIGEON
Circle J Farm   Phone: 269-483-7898
William & Laura Jungjohan
9620 Barker Road [49099] **Directions:** I-80/90 (Indiana Toll Road), Exit 101, head north on Ind. 15, turns into Michigan 103 approx. 4 miles. Look for Barker Road sign, turn right, fourth place on left; look for fencing. **Facilities:** 9 - 12' x 12' stalls, two 12-acre pasture lots, 100' x 200' riding arena, two 50' x 100' pens, feed/hay and trailer parking available. Vet/farrier on call. Reservations preferred. **Rates:** $6 per night; $10 per night with feed. **Accommodations:** Plaza Motel, Tower Motel, Maplecrest Motel in White Pigeon, about 6 miles away.

WILLIAMSTON
Silohetti Manor   Phone: 517-655-3561
Gene Schneider
3725 Norris Road [48895] **Directions:** Located 8 miles from I-96. Call for directions. **Facilities:** 33 indoor 12' x 12' box stalls, 200' x 72' indoor arena, 250' x 100' outdoor arena, feed/hay & trailer parking. Training of hunt seat, Western pleasure, dressage, & jumping done at stable. Call for reservation. **Rates:** $20 per night. **Accommodations:** The Williamston Inn 1/4 mile from stable.

MICHIGAN Page 111

ALL OF OUR STABLES REQUIRE CURRENT NEG. COGGINS, CURRENT HEALTH PAPERS, & OWNERSHIP PAPERS.

WEST BRANCH
✷ <u>Log Haven Bed, Breakfast and Barn</u>　　　　Phone: 989-685-3527
Gail & Paul Gotter　　　　E-mail: gotter@m33access.com
1550 McGregor Rd [48661] **Directions:** I-75 exit 212 into West Branch. Second traffic light, right onto Fairview Rd. North 14 miles, then 1 3/4 miles east on McGregor Rd. Also accessible from M-33 Rose City. **Facilities:** 3 12x12 indoor box stalls, 1 12x12 outdoor stall, 48'x36' turnout, 1 and 1/2 acre pasture, feed/hay available, trailer parking, adjacent to State and Federal Forest with miles of riding trails. **Rates:** $15 per night. **Accommodations:** B&B on site-Call for rates. Motels in West Branch(15 miles) and Rose City(8miles)

YPSILANTI
<u>Sandy Hills Farm</u>　　　　Phone: 734-485-3939
Tom Rudnicki
9101 Cherry Hill [48198] **Directions:** Take Geddes Exit off of US 23. Call for easy directions. **Facilities:** 50 indoor stalls, large indoor & outdoor arenas, pastures, monthly boarding & training available, feed/hay included & trailer parking available. Breeding program & horses for sale. Standing-at-stud: "Cabin Bar Command," AQHA. Tack shop on premises. **Rates:** $20 per night; weekly rate negotiable. **Accommodations:** Knights Inn & Fairfield Inn 6 miles from stable.

Page 112 MINNESOTA

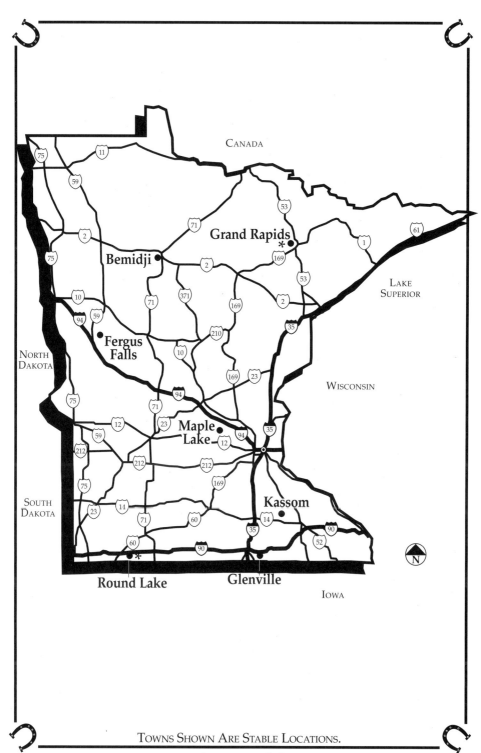

Towns Shown Are Stable Locations.

* onsite accommodations

MINNESOTA Page 113

ALL OF OUR STABLES REQUIRE CURRENT NEG. COGGINS, CURRENT HEALTH PAPERS, & OWNERSHIP PAPERS.

BEMIDJI
Maple Ridge Farm & Stable   Phone: 218-751-8621
Daniel & Peggy Nickerson
3000 Adelia Drive SE [56601] **Directions:** 6 miles off of Hwy 2. Call for directions. **Facilities:** Farm is 80 acres with 12 indoor box stalls, 3 run-in sheds with separate paddocks, 100' x 200' outdoor arena, 6 separate paddocks, large pasture, locked tack room, full lounge, feed/hay & trailer parking available. Arabian stallion standing-at-stud: "Khalim Pasha." Saddlebred stallion standing-at-stud: "Carnival's Loaded Dice." Training of horses & riders at all levels. Call for reservation. **Rates:** $20 per night. **Accommodations:** Edgewater Motel.

FERGUS FALLS
River B Ranch   Phone: 218-736-5134
Brad & Cheri Brause
Rt. 4, Box 109 [56537] **Directions:** Call for directions. **Facilities:** 9 indoor stalls, two 12' x 14' paddocks, three 30' x 80' outdoor paddocks. Will also take cattle. Camper hook-up. Please call for reservation. **Rates:** $15 per night; $75 per week. **Accommodations:** Super 8 is 11 miles from ranch.

GLENVILLE
Margaret & Gaylen Schewe   Phone: 507-448-3786
Rt. 2, Box 103 [56036] **Directions:** 10 miles south of Albert Lea, MN on 35N, Exit 2. Take right off ramp, go 1/8 mile. Take left on County 18 & go 1.5 miles south. Take right on County 83 & go 1 mile. Take right into driveway with red gate. **Facilities:** 3 indoor stalls, 30' x 60' pasture with 8' gates, feed/hay available. Will also take cattle. Call for reservation. Water & electric hook-up for campers. **Rates:** $15 per night. **Accommodations:** Albert Lea on I-90.

GRAND RAPIDS
✷ **K & K Stable, Inc.**   Phone: 218-245-3814
Kathy Horn
1901 Scenic Drive [55744] **Directions:** From US Hwy 2 in Grand Rapids: Go north on State Hwy 38 for 6 miles; go to City Rd 49 & take right; road turns into City 59. Stable is 10 miles down on left. Look for sign. **Facilities:** 6 - 8 indoor box stalls, indoor riding arena, 2 pens with lean-tos which could accommodate 3-6 horses, full-size outdoor arena, pasture/turnout area, lots of riding trails, feed/hay & trailer parking. Camping nearby and on premises. Horses must be wormed within a month. **Rates:** $15 box stall, $10 pen, per night. **Accommodations:** Rustic sleeping lodge available on grounds. 20 minutes from Grand Rapids where there are many motels.

MINNESOTA

ALL OF OUR STABLES REQUIRE CURRENT NEG. COGGINS, CURRENT HEALTH PAPERS, & OWNERSHIP PAPERS.

KASSON
Turn Crest Stables Phone: 507-634-4474
Gene & Vicki Holst
26947 County Hwy. 34 [55944] **Directions:** 10 miles west of Rochester off Hwy 14. Call for directions. **Facilities:** 24 indoor stalls, 14-acre pasture, 4 acres of paddocks, 56' x 152' indoor arena, 90' x 150' outdoor arena. Riding lessons for English hunt seat, jumping, and Western pleasure for all levels. Also horses for sale. Call for reservation. **Rates:** $20 per night. **Accommodations:** Howard Johnson's & Holiday Inn in Rochester, 10 miles from stable.

MAPLE LAKE
Freedom Stables Inc. Phone: 320-963-3351
Kevin & Laura Holen Fax: 320-963-6616
2868 90th Street NW [55358] **Directions:** Exit 183 off of I-94: go south on Co. Rd. 8 for 7.5 miles. Turn east on County Rd. 106 (90th St.) for 2.5 miles. Stable is on left side. **Facilities:** 72 indoor 12' x 11' stalls, 13 paddocks & pastures, 70' x 200' heated indoor arena, 100' x 200' outdoor arena. Hay, grain and bulk bedding available. Trailer friendly parking including semi-truck pull through and loading. 24 hour emergency vet and farriers service available locally. Emergency pick-up can be arranged. Hot and cold water wash stalls, full lounge w/vending. On site camping. Reservations requested, short notice and late night/emergency stopovers OK. Owners reside on premises. Pets welcome. **Rates:** $15 per night; $90 per week. **Accommodations:** Many nearby motels and restaurants. Owners will assist in finding an appropriate facility. Our facility is a favorite stop for national equine transport companies.

ROUND LAKE
* **Painted Prairie Farm** Phone: 507-945-8934
Ralph & Virginia Schenck
RR 1, Box 105 [56167] **Directions:** From I-90: Exit 50 (Hwy 264), Round Lake Exit; go south 4.5 miles. Driveway to farm on left - arch with wagon wheels & brick pillars. **Facilities:** 25 indoor stalls ranging in size from 10' x 10' up to 10' x 20'. Some are wood, some pipe, some with turnout paddocks attached. Paddocks, 30' x 120' runs, & grass pasture available for large groups. Trailer parking & large rigs OK. Alfalfa/grass and good water. Electric camper hook-up available at $5 per night. This farm breeds Paint Horses. Standing-at-stud: "Barlink Tuff Spade" and "Sonny Jet Storm." **Rates:** $15 for stall, $7.50 for paddock per night including feed. **Accommodations:** Prairie House Bed & Breakfast on premises plus motels in Worthington, 10 miles from farm.

MISSISSIPPI

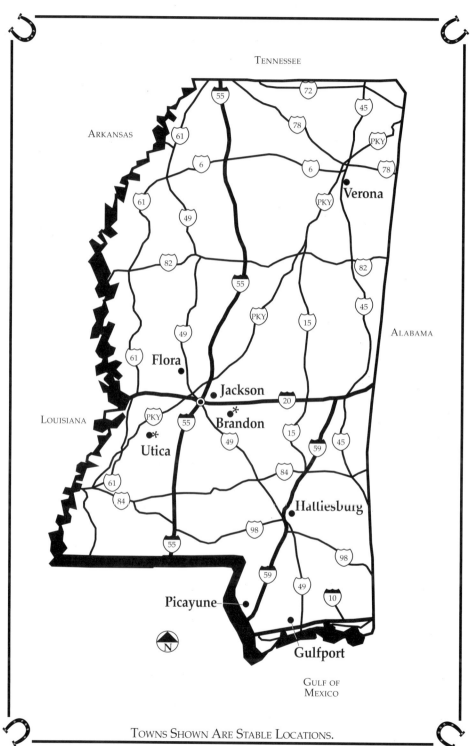

Towns Shown Are Stable Locations.

* onsite accommodations

MISSISSIPPI Page 117

ALL OF OUR STABLES REQUIRE CURRENT NEG. COGGINS, CURRENT HEALTH PAPERS, & OWNERSHIP PAPERS.

BRANDON
* Hilltop Painted Acres Phone: 601-825-2094
John & Barbara Blough   601-940-0246
607 North Street [39042] **Directions:** 12 miles east of Jackson right off of I-20 Exit 59 from East go downtown Brandon to monument turn North at courthose on North St. Follow across RR to farm. **Facilities:** 20 indoor stalls, 25 acres of boarded pasture, 75 acres rental pastures, walker, pipe arena, 200+ acres for riding trails. Breeding, training, & farrier available; veterinarian available. Pens available for rodeo cattle and parking for large rigs. Negative coggins & health papers. Camper hook-ups. Call for reservations. **Rates:** $15 including feeding per night; $60 per week. **Accommodations:** Rustic cabin with barn, $45 per night. Days Inn in Brandon, 3 miles from stable; Comfort Inn in Pearl, 10 miles away.

FLORA
Winterview Farms Phone: 601-879-3468
Paty Nail
140 Spring Road [39071] **Directions:** I-55 to Exit 108, west on Hwy 463; go 3 miles & take left on Robinson Spring Road; go 4.5 miles & turn right on Robinson Spring Road Extension; go 1 mile & turn left on Spring Road. **Facilities:** 10 indoor stalls, 3 1-acre paddocks, round pen, cross-country course, feed/hay & trailer parking available. Fox hunting and polo ponies available. **Rates:** $20 per night. **Accommodations:** Comfort Inn at Ridgeland, 10 miles from stable.

GULFPORT
Harrison County Fairgrounds Phone: 228-832-0080
Mike McMillan
15321 Countfarm Road [39503] **Directions:** I-10 west to Exit 28, north 7.5 miles. **Facilities:** 96 - 9' x 9' wood stalls, trailer parking available. **Rates:** $10 per night. **Accommodations:** Holiday Inn and Best Western at Gulfport, 12 miles from fairgrounds on Hwy 49.

Shady Oaks Stables Phone: 228-832-0435
Ronnie Bourgeois
12726 Wolf River Road [39503] **Directions:** From I-10: Take Exit 28N and go 2 miles to 4-way stop; take left; go 1/4 mile and take right; stable is 2.25 miles on right. **Facilities:** 8 indoor stalls, 20 1-acre fields, 60' x 125' working pen, feed/hay & trailer parking available. 2 full camper hook-ups @ $13.95 per night. **Rates:** $15 per night. **Accommodations:** Many motels 5 miles from stable.

MISSISSIPPI

ALL OF OUR STABLES REQUIRE CURRENT NEG. COGGINS, CURRENT HEALTH PAPERS, & OWNERSHIP PAPERS.

HATTIESBURG
Boots & Saddles Stables　　　　　　　　　　　　Phone: 601-583-6726
Larry Mills
2782 Old Richton Road [39465] **Directions:** Call for directions. **Facilities:** 12 indoor stalls, lighted arena, show equipment, pasture/turnout, trailer parking and feed/hay available. Call for reservations. **Rates:** $7 per night; weekly rate available. **Accommodations:** Motel 6 & Holiday Inn in Hattiesburg, 8 miles from stable.

JACKSON
High Point Farm　　　　　　　　　　　　　　Phone: 601-362-5345
Marcie Lockett
1235 Stigger Road [39209] **Directions:** 9 miles from I-220; 15 miles from I-20 & 10 miles from I-55. Call for directions. **Facilities:** 26 indoor stalls, 4 outside paddocks, 2 outside rings, jump course, 24-hour security. Trains hunter/jumpers at all levels and sells horses. Call for reservations. **Rates:** $20 per night. **Accommodations:** Motels 10 miles from stable.

Mississippi State Fairgrounds　　　　　　　　　Phone: 601-961-4000
Mike Brinkley, Assistant Director
1207 Mississippi Street [39202] **Directions:** High Street exit off I-55, 2 miles north of I-20 and I-55. **Facilities:** 750 indoor 10' x 10' stalls, feed/hay & trailer parking available. Fairground hosts numerous horse activities and events, over 40 per year. **Rates:** $5 per night. **Accommodations:** Ramada Inn, Wilson Inn, Red Roof Inn within 30 minutes.

PICAYUNE
Circle M Riding Stables　　　　　　　　　　　Phone: 601-798-7677
David Megehee
24 Circle M Lane (off Liberty) [39466] **Directions:** Exit 6 off of I-59. In Picayune take Hwy 43 North for 3 miles. Sign on right. **Facilities:** 2 indoor stalls, feed/hay & trailer parking. Overnight boarding in emergency situations only. Please call first. **Rates:** $18 per night. **Accommodations:** Heritage Inn & Majestic Inn 3 miles away on I-59.

MISSISSIPPI Page 119

<u>*ALL OF OUR STABLES REQUIRE CURRENT NEG. COGGINS, CURRENT HEALTH PAPERS, & OWNERSHIP PAPERS.*</u>

UTICA

✶ <u>Big Sand Campground, Inc.</u>   Phone: 601-535-7961
3412 Reedtown Road [39175] **Directions:** From Vicksburg: I-20 take exit 1-c, go straight for 19.2 miles to 4 way stop. Turn right and, go two miles and turn right onto Ross Road. Go 1/2 m to the second drive on the left. From Jackson: Natchez Trace Parkway, take second Utica exit (right hand turn). Go to stop sign, and turn left. Go one mile to 4 way stop, and turn left. Go 2 miles, and turn left onto Ross Road. Go 1/2 mile to the second drive on the left. From Crystal Springs: I-55, take exit 72 (Hwy 27) to Utica. At stop sign, go straight to downtown Utica and 3 way stop. Turn left, go 9.2 miles to 4 way stop, then turn left onto Ross Road. From Port Gibson: Natchez Trace Parkway to Utica exit (first exit past Rock Springs, approx. 5-7 miles) Take exit to first stop sign, and turn left. Go to 4 way stop and turn left onto Ross Road. Go 1/2 mile to the second road on left. **Facilities:** 6 12x12, 12x14 indoor stables, 150x75 turnout, trailer parking available, RV hookup, Pipe Arena. Borders Natchez Trace National Scenic Trail (Horseback & Hiking Only). Negative Coggins **Rates:** $10 per night. **Accommodations:** Bunkhouse in Barn with hot showers. Vicksburg, MS (19.2 miles) Fairfield Inn-Days Inn-EconoLodge.

VERONA/TUPELO
<u>Lee County Agri-Center</u>   Phone: 662-566-5600
5395 Hwy 145 South [38879] **Directions:** Located 15 miles from Hwy 78. Follow signs on Hwy. 45 to the Agri-Center or call. **Facilities:** 110 permanent indoor 10' x 10' box stalls, 53 temporary show stalls, corrals available if notified in advance, 138' x 240' heated indoor arena, lockable stalls. Facility is on 100 acres and has 17 acres of parking. Hours: 7:30 A.M. - 4:30 P.M. Call in advance for stall reservation. 35 RV camper hook-up with water & electricity. RV dump station on grounds. $10 per day. $20 per hour for 200' x 95' outdoor arena. **Rates:** $10 per night. **Accommodations:** Town House Motel in Tupelo, 6 miles from stable.

Page 120 MISSOURI

Towns Shown Are Stable Locations.

* onsite accommodations

MISSOURI Page 121

ALL OF OUR STABLES REQUIRE CURRENT NEG. COGGINS, CURRENT HEALTH PAPERS, & OWNERSHIP PAPERS.

BOONVILLE
✱ <u>Littles Four Oaks Farm</u>　　　　　　　　　　Phone: 660-882-8048
John & Kathy Little
22045 Boonville Rd [65233]　**Directions:** I-70 to SR 87 (Exit 106), south on SR87 for 1.8 mi., left on Boonville Rd. (gravel). Stable 2.4 mi. on right. **Facilities:** 10 indoor 12' x 12' stalls/c paddocks, 10 acres of pasture/turnout, round pen, trailer parking and campers welcome. Pasture riding on 25 acres. Call for reservations. **Rates:** $20/per horse; $125/week. **Accommodations:** B&B on premises; Motels at Boonville, 10 miles from stable.

BOURBON
✱ <u>BourbonMeramec Farm Cabins &</u>　　　　Phone: 573-732-4765
　<u>Trail Riding Vacations LLC</u>　　　　　　　or: 573-732-3080
Carol Springer & David Curtis
E-mail: mfarmbnb@fidnet.com　　　　Web: www.meramecfarm.com
208 Thickety Ford Road [65441]Directions: Exit #218 via I-44 south 9 miles on Hwy "N". Cross Meramec River. Continue 1/4 miles "N" to Thickety Ford Rd make a left, 1/2 mile to "Missouri Century Farm" left again. Cabins, corrals on left. Please call ahead. **Facilities:** 1 indoor, 9 outdoor stalls, trailer parking available, pasture/turnout 48' x 48' corrals. Vacation destination 100's of miles of trails on private & public lands. Breeders of registered MO Fox Trotters. 1 hr - 7 days Inn to Inn trips by reservation or "BYOH" and ride guided or unguided. **Rates:** $10 per night, $50 per week. **Accommodations:** 2 private cabins on premises w/ kitchen, bath, AC/heat. Motels 9-15 miles from farm. Budget Inn, Bourbon. Baymont, Executive Inn, Sullivan, MO.

CARTHAGE
<u>Royalty Arena</u>　　　　　　　　　　　　Phone: 417-548-7722
Mike Pickard　　　　　　　　　　　　　　or 417-358-7711
9895 Cork Lane [64836]　**Directions:** I-44 & Country Rd. #10, Exit 22: Go 1/2 mile north. **Facilities:** 200 inside stalls, 10 outside pens, 2 small grass lots, large parking lot & 20 camper hook-ups. Shavings, feed/hay available. **Rates:** $10 per night.
Accommodations: Days Inn in Carthage, 8 miles from arena; Tara Motel in Joplin, 6 miles from arena, Holiday Inn, Joplin Mo. $43.00/night.

CHESTERFIELD
<u>J. M. Pierce Stables</u>　　　　　　　　　　Phone: 314-394-4733
J. M. Pierce
2315 Baxter Road [63017]　**Directions:** I-64 (Hwy 40) to Clarkson Road south to Baxter Road. Stable is 2 miles on right. **Facilities:** 3 indoor stalls, pasture/turnout, feed/hay & trailer parking available. **Rates:** $20 per night. **Accommodations:** Doubletree & Residence Inn both in Chesterfield, 2 miles from stable.

PAGE 122 **MISSOURI**

<u>*ALL OF OUR STABLES REQUIRE CURRENT NEG. COGGINS, CURRENT HEALTH PAPERS, & OWNERSHIP PAPERS.*</u>

CHILLICOTHE
<u>Indian Creek Equine Center</u>   Phone: 660-646-6227
William "Bill" Hinkebein   E-mail: BHink@greenhills.net
Route 4 [64601]   **Directions:** Located 12 miles from town, north on Hwy 65 to Rte. 190 W, west 9 miles, north 3.5 miles of gravel road. **Facilities:** 3 indoor 10' x10' stalls, 24 stalls at nearby fairgrounds, fenced pastures, feed/hay, trailer parking available. Fox trotters, stallion standing-at-stud: "Hickory's Country Gold," top in nation. North American Trail Ride Conference Competitive Trail Ride. Horses for sale. **Rates:** $20 per night; weekly rates available. **Accommodations:** Motels 10 miles away in Chillicothe.

COLUMBIA
<u>Midway Expo Center</u>   Phone: 573-445-8338
C.W. Adams
I 70 Hwy 40 [65202]   **Directions:** Exit 121 off of I-70. 2 hours east of Kansas City & 2 hours west of St. Louis. **Facilities:** 420 indoor 10' x 10' stalls, 4 indoor arenas (100' x 200' and 50' x 80'), outdoor arena 300' x 150', outdoor riding area. Horse shows & sales held here. Travel plaza with gas & diesel, tire & auto repair, 24-hr restaurant & convenience store. **Rates:** $20 per night. **Accommodations:** Budget Inn on site as well as electric & water hook-ups for RVs - no sewer.

<u>Rangeline Stables & Training Center</u>   Phone: 573-474-0018
Casey and Sandi O'Bryan   Fax: 573-474-0999
E-mail: rangelinestables@centurytel.net
24 S. Rangeline Rd [65201] **Directions:** I-70 Exit 133, 1/2 mile south, go left at stop sign, go 1/4 mile to entrance. Only 3/4 mile off I-70. **Facilities:** 23 12X12 indoor stalls plus indoor arena. Feed/ hay available, trailer parking available. Training facility, farrier services, custom saddles. Negative coggins and shot records required. **Rates:** $25 per night includes bedding. **Accommodations:** Super 8 motel two miles. Exit 131.

CUBA
<u>Blue Moon RV Park & Horse Motel</u>   Phone: 573-885-3622
Liz Barton   Toll Free: 877-440-CAMP
355 Highway F [65453]   Fax: 573-885-3752
Directions: I-44 to exit 203 N on Hwy F, 1500 ft. Paved access. **Facilities:** 6 stalls in covered barn, 1 stall open. 3- 12' x 10' stalls, 3- 12' x 12' stalls. Feed and hay available. Round pen 50', arena 60' x 84'. Blue Moon Equestrian School with lessons and training. All types of livestock accepted. **Rates:** $15 per night, $75 per week. **Accommodations:** Motel 8 in Cuba (5miles), Comfort Inn in St. James (8miles)

MISSOURI Page 123

ALL OF OUR STABLES REQUIRE CURRENT NEG. COGGINS, CURRENT HEALTH PAPERS, & OWNERSHIP PAPERS.

CUBA
Rex Bell Horse Ranch   Phone: 314-623-1310
Rex Bell   E-mail: rexbell@fidnet.com
24265 Vineyard (65453) **Directions:** I-44 Exit 203 – 1/4 mile West on the North Service Rd. **Facilities:** 52 – 12x12 inside stalls. Feed/hay and trailer parking available. Very large indoor and out door arena. Extremely nice inside facilities. Visible from I-44 at 203-mile marker. **Rates:** $20 per night. **Accommodations:** Cuba Super 8, Holiday Inn Express.

DIAMOND
Pear Tree Lane Stables   Phone: 417-325-4136
Harold & Jayne Haskins   E-mail: acofd@jscomm.net   417-437-2191
20160 HWY J [64840] **Directions:** Take exit 18A off I-44-5 miles South to "J" Hwy-2.2 miles. **Facilities:** 23 10x10 stalls, feed/hay available, trailer parking, water & electric hookups, outdoor pens, 1 acre and 1/2 acre pastures, Indoor-Outdoor arena, kennels for cats and dogs. Veterinarian on premisis. **Rates:** $15 per night. **Accommodations:** Motels in Crathage or Joplin. (10 miles).

KANSAS CITY
Benjamin's Ranch   Phone: 816-761-5055
Bob Faulkner, Operator   Web: www.benjranch.com   Fax: 816-761-7400
E-mail: benjranch@aol.com
6401 E. 87th Street, [64138] **Directions:** Located in S.E. Kansas City, two blocks east of I-435 on 87th Street. Call for directions. **Facilities:** 24-12' x 12' indoor stalls with 20' x 20' turnout, pens, arena, isolation area. Vet. available & security 24 hours a day. Complete horse care available. Trailer parking available, $10 electrical hook-up **Rates:** $20 per night. **Accommodations:** Day's Inn across from stable (816-765-4331).

KEARNEY
✱ Over the Hill Ranch, Inc.   Phone: 816-628-5686
Bill & Connie Green   Web: www.othehill.netfirms.com
E-mail: othehill@localnet.com
P.O. Box 743 [64060] **Directions:** 20 minutes North of Kansas City, off I-35. Call for directions. **Facilities:** Eight extra large stalls, Safe and Clean, 60' x 200' indoor arena, 24-hr security. Owner is professional farrier; vet and feed store close by. Jesse James hometown, historical James Farm 7 miles away. Watkins Mill & Smithville Lake nearby. Call for reservations. **Rates:** $20 per night. **Accommodations:** One electrical hook-up for campers, or make reservations at our Western Way B&B. It's the nicest place in the area and a regular stop for many people who have ever stayed once.

MISSOURI

ALL OF OUR STABLES REQUIRE CURRENT NEG. COGGINS, CURRENT HEALTH PAPERS, & OWNERSHIP PAPERS.

LABADIE
Rhodes Riverside Ranch and Equestrian Center   Phone: 636-451-5384
Keith & Julie Rhodes, Owners   Web: www.rhodesriversideranch.com
112 Riverside Drive [63055]   **Directions:** Please visit our website for directions, photo tour and more info about RRR. **Facilities:** 8 12x12 rubber-matted indoor stalls with overhead fans, comfort control heating in winter, 18x30 chat runs, 48x180 grass runs(weather permitting), indoor and outdoor arenas, hay/grass available, trailer parking with electric hookup for $15 per night, wash racks, jump field, ranch trails, Guest Lounge, restrooms/shower. fly/rhino, E&W incef, strangles, west nile, and tetanus required. We only accept cash or check with a valid drivers license. **Rates:** $25 per night. We require a non-refundable $20 deposit per stall 7 working days prior to your arrival. **Accommodations:** Motels, restaurants, and stores within 7 miles of stable.

MARSHFIELD
Cliff Hartman Farm   Phone: 417-859-2200
Mary Chris Hartman
350 Quarter Horse Drive [65706]   **Directions:** Northview Exit off of I-44. Call for directions. **Facilities:** 35 indoor pipe box stalls, indoor arena, round pen, 50 acres of pasture on a total of 450 acres. Farm has 4 superior quarter horse stallions. **Rates:** $25 per night. **Accommodations:** Motels 7 miles away.

✱ **Stevens Farm Inn Horse & Rider**   Phone: 417-859-6525
Walter & Sharon Stevens   E-mail: pony1459@aol.com
5484 State Hwy 00 [65706]   Web: www.usipp.com/stevensfarm
Directions: Less than .5 mile off I-44. From Exit 96, go south on B, east on 00, first place on left. **Facilities:** 8 indoor 8' x 10' box stalls, indoor arena, feed/hay and trailer parking available. Farm is in the Paint and Quarter horse business. Farm faces old historic Route 66. Call for reservations, especially for Saturday. **Rates:** $15 per night per house. **Accommodations:** Suite in barn overlooking arena accommodates 4; house accommodates 4. Small motel within 5 miles. Breakfast by request.

MOSCOW MILLS
Shenandoah Stables   Phone: 636-356-9205
David & Debra Young
116 Majestic Lane [63362] **Directions:** 10 miles north from I-70 or 60 miles south of Hwy 54 & 61 from Hanibal, MO. Right on Hwy 61. **Facilities:** 30 - 10' X 12' indoor wood stalls. Hay, feed and trailer parking available. 2 acre turnout lot. 100' X 200' indoor riding arena, 150' X 350' lighted outdoor arena. Restrooms, snack bar and phone. **Rates:** $15.00 per night $70.00 per week. **Accommodations:** Oak Grove Inn (Troy) 3 miles North of stable. 1-800-435-7144 (special rate if horse boarded with us).

MISSOURI PAGE 125

<u>ALL OF OUR STABLES REQUIRE CURRENT NEG. COGGINS, CURRENT HEALTH PAPERS, & OWNERSHIP PAPERS.</u>

PINEVILLE
<u>Ponderosa Trails</u> Phone: 888-644-6773
Bert & Jean Pekul
1305 Ponderosa Road [64856] **Directions:** Call for Directions **Facilities:** 150 9x10, 9x20 outdoor covered stalls, feed/hay available, trailer parking, 3 110'x300' turnouts, water and electric RV sites, dumpstation, shower house/bathroom, rental units, bedding for stalls, wash racks. **Rates:** $11 per stay up to 7 days. **Accommodations:** Ponderosa Motel (5 mile)

RAYMONDVILLE
* <u>Golden Hills Trail Rides & Resort</u> Phone: 800-874-1157
Charles Golden, Owner or: 417-457-6222
Charlotte Golden-Gray, Contact
19546 Golden Drive [65555] **Directions:** Off Hwy 63 in Houston, MO take Hwy B to Raymondville. Go thru town. 3/4 mile away are signs for High Point Dr. & "Golden Hills Trail Rides." Follow signs. 2 miles off blacktop. **Facilities:** 600 covered stalls with central lighting and water nearby, round pen, arena. Riding trails. Tack store on premises. Minimum of 12-hr notice required. **Rates:** $10 per night. **Accommodations:** 2 large bunkhouses on premises; restroom/shower facilities, 21-room motel: $35 per person per night, $5 each add'l person. Campgrounds on site: $10/night including electrical hook-up.

ST. LOUIS
* <u>The Sleepy P</u> Phone: 618-659-1051
John & Victoria Piel Web: www.sleepy-p.com
E-mail: sleepyp@spiff.net
6309 Miller Drive, Edwardsville, IL [62025] **Directions:** Call for directions. **Facilities:** 10 Stalls and pastures/turnout. All amenities to comfortably accommodate your equine traveler. Feed/hay and vet/farrier on call. **Rates:** Call for rates. **Accommodations:** Bed (comfortable accommodations). Camper/RV hookup and primitive camping. Scenic area. See St. Louis, it has it all! Excellent Indian Museum nearby.

<u>Towne & Country Stables, Inc.</u> Phone: 636-391-7896
Louise Shapleigh
527 Weidman Road, Ballwin [63011] **Directions:** From I-70, take Exit 210A, take Hwy 40-61 towards Kirkwood, go approx. 25 miles to I-270 South, go 1 mile to first exit, west on Manchester Rd. for 2 miles, north on Weidman Rd., 1/2 mile to stable on west side. **Facilities:** Box stalls (limited number available), holding pens, indoor & outdoor arenas, camper hook-up. Reservation required; do not attempt to go thru electric gate — get assistance. **Rates:** $20 per night; discount for 3 or more; $85 per week. **Accommodations:** Several hotels within 15 minutes of stable.

MISSOURI

<u>*ALL OF OUR STABLES REQUIRE CURRENT NEG. COGGINS, CURRENT HEALTH PAPERS, & OWNERSHIP PAPERS.*</u>

ST. LOUIS
<u>Myres Farm</u>   Phone: 573-774-3116
Coby Myres
22945 Reporter Rd [65583] **Directions:** I-44 exit on 156. Go north/west on Hwy to stop light. Turn left at stop light go 1/2 mile turn right on Hwy. Go 2 miles and turn right on Reporter Rd sign. Barn is on left. **Facilities:** 12 indoor stalls. Horse training, starting trail and ranch horses. **Rates:** $15 per night $75 per week. **Accommodations:** Budget Inn, Comfort Inn, Motel 6, Holiday Inn, all 4 miles from stable.

<u>Gateway Stables</u>
See page 75
Pontoon Beach, Illinois

NOTES AND REMINDERS

Page 128 MONTANA

Towns Shown Are Stable Locations.

* onsite accommodations

ALL OF OUR STABLES REQUIRE CURRENT NEG. COGGINS, CURRENT HEALTH PAPERS, & OWNERSHIP PAPERS.

DJ Bar Ranch
5155 Rnd Mtn Rd.
Belgrade, Mt. 59714

(406) 388-7483
cell
(406) 581-7443
info@djbarranch.com
www.djbarranch.com

Montana Horseback Riding Vacation
Many trail heads within easy driving distance, including Yellowstone National Park

Overnight accommodations includes 3 stalls with runs, many 36X48 pens, 100X200 arena & 100X100 arenas
50 foot round pen, $15 stalls or $10 corrals
Water and power hook-ups or a bunkhouse
20 minutes for I 90 between Manhattan - Belgrade
Mules for sale Standing Mammoth Jacks

BELGRADE
DJ Bar Ranch
Jehnet Carlson
E-mail: info@djbarranch.com
Phone: 406-388-7463
Cell: 406-581-7443
Web: www.djbarranch.com
5155 Round Mountain Road [59714] **Directions:** 13 miles from Belgrade. Call for directions. **Facilities:** 320-acre farm/ranch, 3 indoor stalls with runs, 100'x100' arenas, 50' round pen, many 36'x48 corrals, feed/hay & trailer parking on premises, Mules for sale, Standing Mammoth Jack. Montana horseback riding vacation with many trails heads within easy driving distance. **Rates:** $15 per stall, $10 per corral. **Accommodations:** We have water and power hook-ups, or a bunkhouse 20 minutes from I-90 between Belgrade/ Manhattan. E-mail ahead for a map.

MONTANA

ALL OF OUR STABLES REQUIRE CURRENT NEG. COGGINS, CURRENT HEALTH PAPERS, & OWNERSHIP PAPERS.

BIG TIMBER
✱ **Carriage House Ranch** Phone: 406-932-5339
John Haller & Sally DeStefano   Web: www.carriagehouseranch.com
E-mail: chr@carriagehouseranch.com
771 Hwy 191 North [59011] **Directions:** Take Big Timber exit of I-90, follow signs to hwy 191 North, Ranch is 7/10 of a mile past 7 mile marker, on the left. **Facilities:** 10 12x16 indoor stalls with runout, outdoor pens, feed/hay available, trailer parking with electric hookup, 60' round corral, several large paddocks, large indoor arena with excellent footing, outdoor arenas, jumps, CDE (Carriage Driving Course) horse back and carriage trails on 700 acres of spectacular views near the Crazy Mountain trail head. Prefer horse to have strangles and West Nile Virus innoculations. **Accommodations:** Carriage House Ranch B&B (1 mile), Super 8 and Budget Host (8 miles).

BILLINGS
Rickochet Ranch   Phone: 406-259-2011
Rick & T.C.   Cell: 406-690-6072
Web: www.equinemotel.net   Or: 406-690-6079
E-mail: rickocheranch@hotmail.com
3306 Becraft Ln. [59101] **Directions:** East side of Billings, close to I-90. Exit 455 Johnson Ln. Go east for 0.2 miles toward the Flying J, turn left on Old Harding Rd. Go northeast for 300' turn right on Becraft Ln. Go east for 0.6 miles to 3306 Becraft. **Facilities:** 14 – 12X20 indoor stalls, 18 – 12X60 outside paddocks w/ cover, 5 – 16 X 48 pens with cover, all stalls and pens have fresh continuous running water, room for big rigs & trailer parking on premises. Indoor wash-stall w/ warm water, mare foaling stall, round pen, outdoor arena, trails, training, lessons and timothy grass hay available. Fair grounds 4 miles away. **Rates:** $15 with out hay, $18 with hay. **Accommodations:** Many motels in Billings.

June's Horse Motel   Phone: 406-252-9563
June Nagel   or: 406-248-4944
406 Johnson Lane [59101] **Directions:** Exit 455 off I-90, .5 mile south on Johnson Lane. **Facilities:** 2 indoor stalls, 3 stalls with 18' x 30' outside runs, corral at barn and working round pen, hay grown on premises, sawdust for bedding, trailer parking based on size and weather, security camera. Call for reservations. **Rates:** $15 per night. **Accommodations:** Motels within 4 miles.

BOZEMAN
✱ **Gallatin River Lodge**   Phone: 406-388-0148
Steve & Christy Gamble   or: 406-388-9435
Web: www.grlodge.com
9105 Thorpe Road [59718] **Directions:** 3 miles south of I-90(exit 298- Belgrade) **Facilities:** 15 indoor/outdoor stalls, feed/hay available, trailer parking, 20-acre turnout, Please Call Ahead! **Rates:** $15-20 per night, $100 per week. multiple horse discounts **Accommodations:** Lodging for "people" also available with an excellent restaurant on site, Holiday Inn Express and Super 8 -3 miles away.

MONTANA   Page 131

ALL OF OUR STABLES REQUIRE CURRENT NEG. COGGINS, CURRENT HEALTH PAPERS, & OWNERSHIP PAPERS.

BUTTE
No Excuses Arena   Phone: 406-782-4540
Paula Scott & Audrey Chamberlin   Arena: 406-494-6547
21 Elgin Drive- Elk Park [59701]   Cell: 406-490-6000
Directions: 2 miles N of junction I-15 and I-90, 5 miles N of Butte, visible from I-15 (call for specific directions). **Facilities:** 19 indoor, matted box stalls, outdoor runs, pasture. Feed/hay available and 100 x 200 arena. **Rates:** $20 per night indoor stalls. **Accommodations:** B & B, Motels 5 miles.

DEER LODGE
Mountain View Arena   Phone: 406-846-1989
Alex & Kayo Fraser   E-mail: info@drivehorses.com
Web: www.drivehorses.com & www.wildhorsebooks.com
255 Boulder Road [59722] **Directions:** You can see the arena from I-90 Westbound traffic; take exit #184, turn right and take the first right. We are the second drive on left, next to the Vet clinic. Eastbound traffic, take exit #184, turn left at stop sign, go under the freeway, take the first right. We are the second drive way on the left. **Facilities:** 30 indoor stalls 12'x12', no pasture or runs. Outdoor pens are for full time boarders not over night horses. 70'x 230' indoor arena. Hay and trailer parking available on premises. Driving lessons, training and clinics. Equine books and art for sale on premises. We request visiting horse be current on all vaccinations. **Rates:** $15 per night, extra shavings for sale. Reservations appreciated but not required. **Accommodations:** Motels, restaurants and campgrounds across the freeway from us. Museums and Grant/Kohrs Ranch National Park close by. This is a good place to rest over for a couple of days.

FLORENCE
✱ Parsons' Pony Farm   Phone: 406-273-3363
Suzi Kimzey Parsons
5710 Yarrow Drive [59833] **Directions:** West of Florence on Hwy 93. Call for directions. **Facilities:** 3 enclosed stalls, 3 open stalls with access to paddocks, 3 pastures from 1 to 3 acres each, feed/hay and trailer parking available. Camper hookup. Beautiful extensive trails nearby in Bitterroot Mts., dog kennels available. Cart & saddle ponies, lessons, training, trail rides just for kids. B&B on premises, tennis court. **Rates:** $10-$12 per night. **Accommodations:** B&B on premises; 2 guest bedrooms with shared bath. Days Inn in Lolo, 10 miles from farm.

GALLATIN GATEWAY
✱ Wild Rose Bed & Breakfast   Phone: 406-763-4692
Dennis & Diane Bauer
1285 Upper Tom Burke Road [59730] **Directions:** Call for directions. Located 2 miles south of Gallatin Gateway. **Facilities:** 2 corrals (one large, one small), both with water, 3-5 acres of pasture/turnout, feed/hay and trailer parking available. **Rates:** $15 per night. **Accommodations:** B&B on premises. Other accommodations 2-15 miles from stable.

MONTANA

ALL OF OUR STABLES REQUIRE CURRENT NEG. COGGINS, CURRENT HEALTH PAPERS, & OWNERSHIP PAPERS.

GREAT FALLS

B – C Stables   Phone: 406-761-7426
Keith Lewis
33 - 60th Street South [59405] **Directions:** 5 miles from I-15 and 4 blocks from by-pass. Call for directions. **Facilities:** 33 indoor stalls, heated auto waterer, 120' x 180' outdoor arena, paddocks, breaking arena, feed/hay & trailer parking on premises. Call for reservation. **Rates:** $10 per night. **Accommodations:** Highwood Motel 1 mile from stable.

Lazy D Arena &Stables   Phone: 406-964-0733
David & Caroline Knudson   E-mail: LazyDarena@hotmail.com
200 US HWY 89 [59487] **Directions:** 2 miles off Interstate 15 on Hwy 89. **Facilities:** 35 indoor stalls, 50 outdoor stalls. Outdoor runs and paddocks. Lazy D Arena sits on 200 acres. Outdoor arena is 150' x 300'. Open riding $15 per person-no time limit. Indoor arena 110' x 260' both have excellent ground. We offer riding membership, overnight, weekly and monthly boarding. Hay, feed and shavings are available upon request. All event facility. Riding lessons available. Clean coggins and health certificate with papers must be available to board. **Rates:** $15 per night, $65 per week. **Accommodations:** Nearby hotels are within 10-15 mile radius, Hawthorne, Town House, Crystal Inn.

Skyline Horse Hotel & Vet Clinic   Phone: 406-761-8282
Nora Seekins/ Sara & John Seekins   Fax: 406-761-6900
Web: www.vet4yourpet.com   E-mail: nksnpets@aol.com
Junction of Bootlegger Trail & Haver Highway/3700 US Hwy 87 [59403]
Directions: Call for directions. **Facilities:** 7 indoor stalls, 3 outdoor paddocks, and one outdoor arena. Clean facility and accommodating staff. Call for reservations. Vet service available. **Rates:** $20 per night. **Accommodations:** Days Inn 1 mile from clinic (406-727-6565) Best Western Heritage Inn (406-727-7200) *Ask for special Skyline Horse Hotel rates(discount 15%).

HAVRE

The Great Northern Fairgrounds   Phone: 406-265-7121
Mike Spencer
1676 Highway 2 West [59501-6104] **Directions:** Right on Hwy 2 west of town. **Facilities:** 35 metal 10' x 10' indoor stalls, 200' x 140' indoor arena, 180' x 300' outdoor arena, 20 camper/RV hook-ups, trailer parking available. **Rates:** $5 per night; $20 deposit. Must clean stall. **Accommodations:** Campers $5 per night no utilities, $12.40 with utilities.

MONTANA Page 133

ALL OF OUR STABLES REQUIRE CURRENT NEG. COGGINS, CURRENT HEALTH PAPERS, & OWNERSHIP PAPERS.

JORDAN
✱ Sand Creek Clydesdales Ranch Vacations   Phone: 406-557-2865
Wade & Bev Harbaugh   Web: www.sandcreekclydedales.com
E-mail: bev@midrivers.com
Hwy 200 East, P.O. Box 330 [59337] **Directions:** Hwy 200 runs east to west through Montana, ranch is east of Jordan, near mile marker 220 (7 miles out of Jordan) then 3 miles south of hwy. **Facilities:** 3 indoor stalls and 4 outdoor pens, pasture/turnout area. Feed/hay available, trailer parking. Ranch vacations, overnight lodging and wagon trains. **Rates:** $10 per night, $70 per week. **Accommodations:** Lodging available at Ranch for $65 per couple.

LIVINGSTON
Park County Fairgrounds   Phone: 406-222-4185
Kim Knutson, Manager
46 ViewVista Drive [59047] **Directions:** Exits 333 South, or 331 West, or 335 East off I-90. Turn on Park Street and travel to H Street. Turn down H Street onto Vista View Drive. **Facilities:** 22 indoor stalls, 6 open covered stalls, 16 outside stalls, trailer parking. **Rates:** $10 per night. **Accommodations:** Best Western Yellowstone, Paradise Inn, Comfort Inn. Super 8 Motel, EconoLodge all in Livingston.

REED POINT
✱ S Bar K Ranch   Phone: 406-326-2280
Blanche Davis, Joe Davis, Mary Berry
#1 Dead End Road [59069] **Directions:** 2 miles from I-90. Exit 392 off of I-90. Go north through town and across railroad tracks and Yellowstone River Bridge. Continue north until you see a cattle guard, turn right onto Dead End Road, first place on left. **Facilities:** Working cattle ranch. 4 indoor 8' x 10' box stalls, 1 15' x 23' covered barn pen, 3 - 38' x 10' runs with cover, 3 solid-board 16' x 22' box pens, 18' x 46' barn-connected box pen, 2 corrals approx. 500 sq. yds, no turnout/pasture, feed/hay & trailer parking available. 640-acre area for riding. Proof of Influenza, Rhinopneumonitus shots & worming. **Rates:** $10-$20 per night. with hay; $75 per week **Accommodations:** On premises: 2 bedrooms with bath, double beds, & breakfast, $75 per night. Super 8 motels in Columbus (20 miles east) and Big Timber (26 miles west).

SEELEY LAKE
✱ Horseshoe Hills Guest Ranch   Phone: 406-677-2276
Wayne & Rena Heaton
6190 Woodworth Road [59868] **Directions:** Scenic Hwy 83, N. off Hwy 200. At MM 7 take Woodworth Road, 5 miles to ranch entrance. Call for more details. **Facilities:** 17 covered, 10' X 20' outdoor, 18 indoor, 10' X 12' heated stalls. More with temporary panels. Corrals, 40' and up. Pastures 1/2 acre and up. Breeding, boarding & training. Paint and quarterhorses for sale at all times, all levels. Unlimited trails on state & national forest and Bob Marshall Wilderness. **Rates:** $6 to $15 per horse. **Accommodations:** Ranch rooms and camp sites on premises, (no pets in rooms - kennels available), B&B 1 mi., Seeley Lake Motels 20 minutes to town.

MONTANA

ALL OF OUR STABLES REQUIRE CURRENT NEG. COGGINS, CURRENT HEALTH PAPERS, & OWNERSHIP PAPERS.

VICTOR

✽ <u>Bear Creek Lodge</u> Phone: 406-642-3750
Roland & Elizabeth Turney Web: www.bear-creek-lodge.com
1184 Bear Creek Trail [59875] **Directions:** Call for directions. **Facilities:** 3 large indoor stalls, 3 pens, 5 acres pasture adjacent to the Selway-Bitterroot Wilderness (1.6 million acres) with numerous riding trails starting from the property or within a short drive from the lodge. **Rates:** Free to our guests. **Accommodations:** On premises: small secluded, exquisite lodge featuring fine dining; $21 single and $300 double per night with meals.

NOTES AND REMINDERS

Page 136 **NEBRASKA**

Towns Shown Are Stable Locations.

* onsite accommodations

NEBRASKA   Page 137

ALL OF OUR STABLES REQUIRE CURRENT NEG. COGGINS, CURRENT HEALTH PAPERS, & OWNERSHIP PAPERS.

BAYARD
✱ <u>Flying Bee Ranch LLC</u>   Phone: 888-534-2341
Conrad & Louise Kinnaman   Web: www.flyingbee-ranch.com
E-mail: flyingbee@bbc.net
6755 Cty Rd. 42 [69334] Directions: Go 5 miles South of McGrew off Hwy 92 on County Road 34, turn East on County Road 42 for 1/2 mile. Ranch buildings on left. **Facilities:** 18 outdoor pipe stalls, 7 indoor panel stalls, 2 indoor box stalls with runs, 4 large corrals, hay for sale. We are a working cattle ranch with 4,000 acres of trails to ride, guest rooms, cabin, campground and boarding facilities. **Rates:** $8-15 per night. **Accommodations:** Motels available in Scottsbluff, NE (20 miles)

CHADRON
<u>Panhandle Veterinary Clinic</u>   Phone: 308-432-2020
John E. Gamby, DVM
985 Hyway 385 South [69337] Directions: 1/2 mile south of Hwy 20 on Hwy 385. **Facilities:** 2 indoor stalls (10' x 12') 2 outdoor stalls 12' x 12'. **Rates:** $10 per night, $60 per week. **Accommodations:** Best Western (across from Hwy), Days Inn (1/4 mile), Super 8 (1 mile). General veterinary service.

COLUMBUS
<u>Columbus Racetrack/Fairground</u>   Phone: 402-564-0133
Gary Bock
15th Street & 10th Avenue [68601] Directions: From US 30 or US 81, follow signs to racetrack. **Facilities:** 800 indoor stalls, arena, feed/hay & trailer parking available. This is a large thoroughbred racetrack with over 800 stalls. **Rates:** $6 per night. **Accommodations:** Sleep Inn 1/2 mile from track.

CRAWFORD
✱ <u>Ash Creek Ranch Vacations</u>   Phone: 308-665-1580
Gary & Nancy Fisher
Ash Creek, Inc.,
617 West Ash Creek Road [69339] Directions: Call or write for brochure. **Facilities:** 4 indoor stalls, 20 acres of pasture/turnout, feed/hay & trailer parking available. 1,800-acre working ranch, thousands of acres of adjoining National Forest land, beautiful area for trail riding. **Rates:** $10 per night, $60 per week. **Accommodations:** 100-yr.-old ranch house with two bedrooms, kitchen, living room, indoor bathroom. Sleeps up to ten. $75 per night for two; $6 each additional person. No housekeeping provided.

<u>Fort Robinson State Park</u>   Phone: 308-665-2900
Mike Morava
Box 392 [69339] Directions: Located 3 miles west of Crawford on Hwy 20. **Facilities:** 120 indoor stalls, no pasture/turnout area, trailer parking but no feed available. **Rates:** $6 per night. **Accommodations:** Cabins to rent on site from April to November: 2 to 9 bedrooms starting at $54 per night with lodge rooms at $28 to $33 per night. Stables closed November 1 –April 1.

NEBRASKA

ALL OF OUR STABLES REQUIRE CURRENT NEG. COGGINS, CURRENT HEALTH PAPERS, & OWNERSHIP PAPERS.

ELKHORN
Quail Run Horse Center   Phone: 402-289-2159
22021 West Maple Road [68022] **Directions:** Take Maple St. Exit off of I-680. Call for directions. **Facilities:** 45 indoor stalls, arena, feed/hay included, trailer parking available. A teaching, training, & show facility specializing in English & hunter/jumper. Call for reservation & availability. **Rates:** $15 per night. **Accommodations:** Motels in Omaha 10 minutes from stable.

GRETNA
T S Arabians, Inc.   Phone: 402-332-4328
Ted Smalley
22603 Fairview Road [68028] **Directions:** Exit 432 off of I-80. Call for directions. **Facilities:** 47 indoor stalls, round pen, 70' x 176' indoor arena, 4 outdoor runs with pasture, heated & air-conditioned lounge. Breeds, sells, & trains Arabians, quarter horses & thoroughbreds. Riding lessons for all levels. Call for reservation. **Rates:** $15 per night. **Accommodations:** Motels 7 miles from stable.

LEXINGTON
Horse Motel &Puppy Shop   Phone: 308-324-6303
Cyndi Ocken   Cell: 308-324-7650
Web: puppy-shop.com   E-mail: puppyshop@cozadtel.net
204 W. River Road, [68850] **Directions:** 1 mile west of I-80. Call for directions. **Facilities:** 8-12 x 12' indoor stalls, round pen, outdoor arena, trailer parking available. Riding trails to Platte River. Health papers. **Rates:** outside stalls-$10 per head, per night, inside stalls-$20 per head, per night. **Accommodations:** Super 8, Comfort Inn, & Days Inn.

LINCOLN
K/B Stables   Phone: 402-465-5855
Kenneth & Berna Stading
6100 N. 98th [68507] **Directions:** Please call ahead for reservations. Take Waverly Ext. going west. Turn left on 98th & go about 1 mile. **Facilities:** 1 stall plus paddocks and inside arena. Feed & hay available. Trailer parking available. Pasture available. **Rates:** $15.00. **Accommodations:** Quality Inn in Lincoln, 2 miles from stable.

NORFOLK
Norfolk Livestock Market   Phone: 402-371-0500
Ask for Junior
1601 South 1st Street [68701] **Directions:** Take Hwy 81 to Omaha Ave.; go east on Omaha 1 mile & turn right and go 2 blocks past tracks and you will see sign. Also easily accessible from Hwy 275 from Omaha. **Facilities:** 30 indoor stalls, pasture/turnout pens, feed/hay & trailer parking on site. **Rates:** $2 per night. **Accommodations:** Holiday Inn & Super 8 in downtown Norfolk 1 mile from stable.

NEBRASKA   Page 139

ALL OF OUR STABLES REQUIRE CURRENT NEG. COGGINS, CURRENT HEALTH PAPERS, & OWNERSHIP PAPERS.

NORTH PLATTE
Remuda Stables   Phone: 308-532-4359
Ron & Dawn Andersen
Rt 3, Box 257 [69101] **Directions:** Easy access -- Please Call. **Facilities:** Total: 24 stalls. 10 available for overnight use. 14 open front shed, 12' X 13'/w 40' runs. 3 - 22' X 30', 7 - 12' X 12' outside with good protection. Feed, hay and trailer parking available. No pasture available. 150' X 250' arena and 60' round pen. **Rates:** $11 per night per horse. **Accommodations:** 5 miles to restaurants and motels.

OGALLALA
Baltzell Veterinary Hospital   Phone: 308-284-4313
C.W. or David Baltzell   Cell: 308-289-2605
1710 West 4th Street [69153] **Directions:** Located off of I-80. 1/4 mile north of Hwy 30. Call for directions. **Facilities:** 1 indoor, 3 outdoor stalls, 1 acre of fenced pasture/turnout, feed/hay & trailer parking on site. Medical services available at clinic. Call for reservation. **Rates:** $15 per night; group rates & feed available. **Accommodations:** Many motels 1 mile from hospital.

Peterson Stables   Phone: 308-284-8235
K.C. Peterson
851 Rd. West D North, Ogallala [69153] **Directions:** Located off I-80, 4 miles west on Hwy 30, 1/2 miles north. **Facilities:** 20 Box stalls with turnouts, indoor arena, feed, hay and trailer parking available. **Rates:** $12.50 a night per horse.

YORK
Diamond B, Inc.   Phone: 402-362-5439
Bryan & Diane Buss
Rt. 1, Box 154-A [68467-9781] **Directions:** Call for directions. **Facilities:** 3+ box stalls, turnout pens, 70' x 230' indoor arena, 100' x 300' outdoor arena, feed/hay available, trailer parking on site. Camping allowed. Please call evenings or weekends for reservation. **Rates:** $15 per night; $75 per week. **Accommodations:** Staehr Motel in York 3 miles from stable.

Page 140 NEVADA

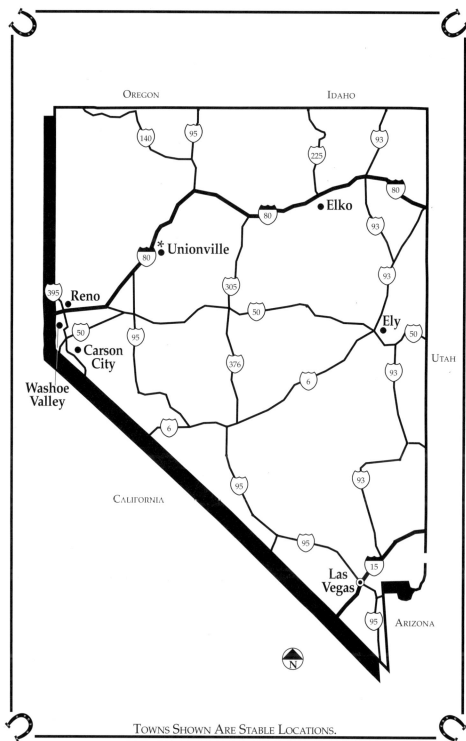

Towns Shown Are Stable Locations.
* onsite accommodations

NEVADA Page 141

ALL OF OUR STABLES REQUIRE CURRENT NEG. COGGINS, CURRENT HEALTH PAPERS, & OWNERSHIP PAPERS.

CARSON CITY
Equest Training Center   Phone: 775-849-0105
Vicki Sherwood
805 Washoe Drive [89704] **Directions:** Located on Hwy 395. Call for directions. **Facilities:** 15 indoor stalls, 14 outdoor paddocks, dressage arena, jumper field, 60' x 100' covered arena, round pen, wash rack. 24-hr care. Training for dressage, jumping & 3-day events at all levels. Overnight boarding in emergencies only. **Rates:** $20 per night; call for discounted weekly rate. **Accommodations:** Motels 1 mile from stable.

Old Washoe Stables   Phone: 775-849-1020
Michael Stockwell
1201 Hwy 395 [89704] **Directions:** Call for directions. **Facilities:** At least one indoor stall, 6 - 12' x 12' holding pens, 10' x 20' indoor corral. Guided horseback rides at $20/hr. and children's riding. Advance reservations required. **Rates:** Will be competitive. **Accommodations:** Round Hill Station in Carson City, 7 miles from stable.

ELKO
Elko County Fairgrounds   Phone: 775-738-7925
Ask for Jeanie or Angelo
P.O. Box 2067 [89803] **Directions:** Located off of I-80. Call for directions. **Facilities:** 365 indoor stalls, arena, 5/8 mile race track. Hay & trailer parking available. Open 24 hours. 1 horse per stall. Check in at double-wide trailer at entrance. **Rates:** $10 per night. **Accommodations:** Motels 1/4 mile away.

Suzie Creek Arabians   Phone: 775-738-8631
PO Box 1360 [89803] **Directions:** I-80, Hunter exit 292, 3 miles south. **Facilities:** 14 indoor stalls, 6 outdoor stalls, holding pens, provide your own hay and feed. Indoor and outdoor arenas, walker and lots of trails. **Rates:** $20 per night indoor, $15 per night outdoor. **Accommodations:** Motels in Elko, 10 miles from stable.

ELY
White Pine County Fairgrounds   Phone: 775-289-4691
Sterling Wines
McGill Highway, Rt. 93 [89315] **Directions:** Call for directions. **Facilities:** 200 inside stalls, 6 outside stalls, no feed/hay. Trailer parking available. Open 24 hours. **Rates:** $10 per night, inside stall; $5 per night, outside stall. **Accommodations:** Motels 2 miles from fairgrounds.

PAGE 142   NEVADA

ALL OF OUR STABLES REQUIRE CURRENT NEG. COGGINS, CURRENT HEALTH PAPERS, & OWNERSHIP PAPERS.

LAS VEGAS
Bamberry Stables   Phone: 702-361-6620
Don Bamberry
7475 Rogers [89139]   **Directions:** Exit 33, I-15 S., Quiet neighborhood 1 mile from I-15. **Facilities:** 12 outdoor pipe corrals 20' x 20' with shade, 100' x 130' arena, trailer parking. **Rates:** $15 per night per horse. Cash only. **Accommodations:** Many hotels and motels close by. Call for additional information.

Miller Ranch   Phone: 702-645-2350
E-mail: hrseluvertoo@aol.com   Fax: 702-645-2509
5837 El Capitan Way [89149]**Directions:** 2 miles from I-95. Call for directions. Facilities: 10 stalls, 3 indoor and 7 outdoor. Feed/hay available, trailer parking available. Arena, wash rack and round pen. Rates: $15 per horse per night. Accommodations: 3 miles from Las Vegas.

RENO
✱ **Meachum Ranch**
See Page 143

UNIONVILLE
Old Pioneer Garden Country Inn   Phone: 775-538-7585
The Jones Family
2805 Unionville Road [89418]   **Directions:** 20 minutes from Exit 149, Unionville Exit off I-80. **Facilities:** 3 indoor 12' x 10' stalls connecting with paddocks, 1/2-acre fenced pasture, feed/hay, trailer parking available. Trail riding, gold mines nearby, Winnemucca has gambling (1 hr away). Only 18 people in this ghost town. **Rates:** Horses free when you stay at the Inn. **Accommodations:** Inn on premises, $75-$95 per night including breakfast, other meals available.

WASHOE VALLEY
Franktown Meadows   Phone: 775-849-1600
Janice
4200 Old Hwy 395 North [89704]   **Directions:** 20 miles south of I-80. Call for directions. **Facilities:** On 41 acres. 63 indoor stalls, 12 pasture/turnout areas, indoor arena, paddock, outside arena, jump course, dressage court. Feed/hay & trailer parking available. Training of hunter/jumper & dressage for horses & riders. Call for reservations. **Rates:** $15 per night. **Accommodations:** Motels 6 miles from stable.

ALL OF OUR STABLES REQUIRE CURRENT NEG. COGGINS, CURRENT HEALTH PAPERS, & OWNERSHIP PAPERS.

RENO
Meachum Ranch
Monty & Michelle Meacham
E-mail: bentley202@aol.com
2150 Greentree Lane [89511]

Phone: 775-851-3456
Fax: 775-851-1098

Directions: US 395 to S. Virginia St. Exit #61 right on S. Virginia St, .37 miles to Holcomb Ranch Lane, turn right. 1.22 miles to Greentree turn right .13 miles to end of street, dead ends into our ranch.
Facilities: 35 stalls. Large indoor & outdoor riding arenas. Most stalls have outdoor runs. Feed/hay available, trailer parking available. **Rates:** Call for rates.
Accommodations: Courtyard by Marriot - 3 miles from ranch (775-851-8300), 5 miles to Atlantis Hotel & Casino. 10 miles to Reno Livestock Event Center.

Meacham Ranch

AMHA REGISTERED HORSES
MONTY & MICHELLE MEACHAM
2150 GREENTREE LANE
RENO, NV 89511
TEL: 775-851-3456   FAX: 775-851-1098

BOARDING FOR ALL BREEDS
BY THE DAY OR BY THE MONTH

- Large Indoor & Outdoor Riding Arenas
- Most Stalls Have Outdoor Runs
- Daily Turnout Available
- Wash Rack with Hot Water
- Parking for Horse Trailers
- 5 Miles to Hotel and Casino
- 10 Miles to Livestock Event Center
- Reservations Recommended

Page 144 NEW HAMPSHIRE

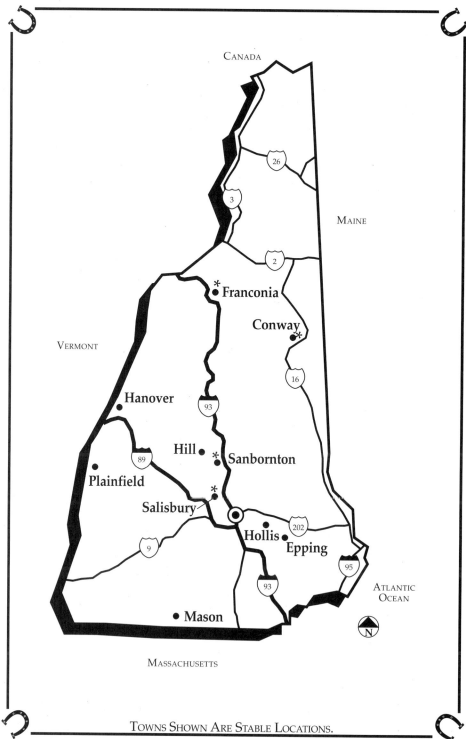

Towns Shown Are Stable Locations.
* onsite accommodations

NEW HAMPSHIRE

ALL OF OUR STABLES REQUIRE CURRENT NEG. COGGINS, CURRENT HEALTH PAPERS, & OWNERSHIP PAPERS.

CONWAY

* The Foothills Farm Bed & Breakfast Phone: 207-935-3799
Theresa & Kevin Early
P.O. Box 1368 [03818] **Directions:** Call for directions. **Facilities:** 3 indoor 10' x 10' stalls, 2 outdoor 12' x 12' stalls, three 1-acre pastures, one turnout area, feed/hay & ample trailer parking available. Quiet riding on marked trails that range from 8 to 22 miles in length - no busy roads to cross. Guided trail rides available. **Rates:** $18 per night. **Accommodations:** 4-bedroom B&B on premises: $48/dbl including full breakfast.

EPPING

Rum Brook Farm Phone: 603-679-5982
Meg Preston
44 Hedding Road (Rte. 87), P.O. Box C [03042-1302] **Directions:** From I-495, take Rte 125 north, east on 87. From I-95, take Rte 101 west to Rte 125 north, right on 87. Farm is 1/3 mile on 87. **Facilities:** 30 indoor 10' x 12' box stalls, large and small paddocks, indoor arena, feed/hay, trailer parking. Breeding facility; standing-at-stud: "Immortal Command" and "Serenity March Time." Morgan horses for sale. Equitation program. **Rates:** $15 per night, $94 per week. **Accommodations:** Best Western in Exeter, 15 miles away; Epping Motel in Epping, 1 mile away; motels in Hampton Beach, 20 miles away.

FRANCONIA

* Bungay Jar Bed and Breakfast Phone: 603-823-7775
Kate Kerivan Toll Free: 1-800-421-0701
P.O. Box 15, Easton Valley Road [03580] Fax: 603-444-0100
Directions: From I-93 North, take Exit 38. South on Hwy 116 for 5.5 miles. Bungay Jar on left just past Sugar Hill Road. From I-91 North, take Exit 17. Take Hwy 302 east through Woodsville for about 12.5 miles. Turn right onto Hwy 117. From Franconia, go south on 116. Bungay Jar is 5.5 miles on left. **Facilities:** Turnout only, no stalls; hay & pasture/turnout available, bring own grain; trailer parking available. 15 acres in White Mountains with National Forest and AMC trails out the back pasture. Dressage, hunter/jumper clinics nearby. **Rates:** $15 per night. **Accommodations:** B&B on site, $75-$150 per night (dbl), higher during foliage season. Country breakfast included.

NEW HAMPSHIRE

ALL OF OUR STABLES REQUIRE CURRENT NEG. COGGINS, CURRENT HEALTH PAPERS, & OWNERSHIP PAPERS.

HANOVER
Velvet Rocks Farm   Phone: 603-643-2025
Marilyn Blodgett
24 Trescott Road [03750] **Directions:** Call for directions. **Facilities:** Up to 6 indoor box stalls with rubber mats, riding ring, no pasture/turnout, feed/hay & trailer parking available. Call for reservations. **Rates:** $15 per night. **Accommodations:** Radisson & Sheraton in White River Junction, about 5 miles from stable.

HILL
Gloria King's Stables & Trail Rides and Boarding   Phone: 603-934-5740
Leon & Gloria King
RR 1, Box 1191 [03243] **Directions:** Exit 23 off of I-93. Call for directions. **Facilities:** 31 indoor stalls, pasture, feed/hay & trailer parking available. Tack store on premises. Acres to ride on plus river crossing. **Rates:** Rides, $20 per hr; $35, two hrs. Boarding: $15 per night; $75 per week; $125 per month. **Accommodations:** B&B 5 miles from stable.

HOLLIS
Glory Hole Ranch   Phone: 603-465-2672
Rich Lasal
Wheeler Road [03049] **Directions:** Only 45 minutes from Boston, call for directions. **Facilities:** 10 quality 12'x12' indoor stalls, 10 acre pasture, feed/hay available, trailer parking. Brand new barn w/indoor & outdoor arenas. 5 minutes from Nashue and near major highways. **Rates:** $20 per day. **Accommodations:** Many hotels/motels in the area.

MASON
Hearthstone Farm   Phone: 603-878-3046
Barbara Baker
610 Greenville Road [03048] **Directions:** Call for directions. **Facilities:** 6 indoor stalls, 60' x 180' indoor ring, 3 paddocks on 55 acres. Buys & sells horses and offers full training & lessons. Parking for big rigs available. Call for reservations. **Rates:** $15 per night. **Accommodations:** B&Bs within 10 miles of stable.

PLAINFIELD
MNMS Stables   Phone: 603-675-2915
Sue Zayatz   NH Residents: 800-871-2915
Westgate Road [03781] **Directions:** Call for directions. **Facilities:** 13 indoor 12' x 12' stalls, 10 outside paddocks, 10-acre pasture, 64' x 120' indoor arena, outside ring, hunter/jumper course. Horse transportation throughout New England. Call for reservations. **Rates:** $15 per night; ask for weekly rate. **Accommodations:** Motels 7 miles from stable.

NEW HAMPSHIRE   PAGE 147

ALL OF OUR STABLES REQUIRE CURRENT NEG. COGGINS,
CURRENT HEALTH PAPERS, & OWNERSHIP PAPERS.

SALISBURY
* **Horse Haven B&B**   Phone: 603-648-2101
Velma Emery
462 Raccoon Hill Road [03268] Directions: I-93 N from Concord. Exit 17 onto Rt 4. 9 miles to Salisbury. Call for details. **Facilities:** Beautiful scenic New Hampshire country side. Thoroughbred foals frolicking on the hillside. 35 acre farm. Horse stalls for B&B guests only. Outdoor 12' X 12' outdoor pipe frame corrals Free. **Rates:** Large indoor bedded stall $15.00 per night. **Accommodations:** Double Occupancy/w shared bath $65.00. Single occupancy/w shared bath, $35.00 to $50.00, Extra person, $15.00. Room rates include complete continental breakfast.

SANBORNTON
* **Tir Na N'og Stables**   Phone: 603-528-5068
Patricia Walsh
35 Parker Hill Rd (03269) Directions: Exit 20 off I-93. Call for directions.
Facilities: 10'x10' matted stalls, round pen, all day turn out, 12.5 pasture, hay/shavings provided. Enjoy miles of trials in the beautiful lakes region. **Rates:** $20 per night, $80 per 5 days. **Accommodations:** B&B offered to horse owners only. Queen bed, built in bed, continental breakfast for $80 double-occupancy per night.

Page 148 NEW JERSEY

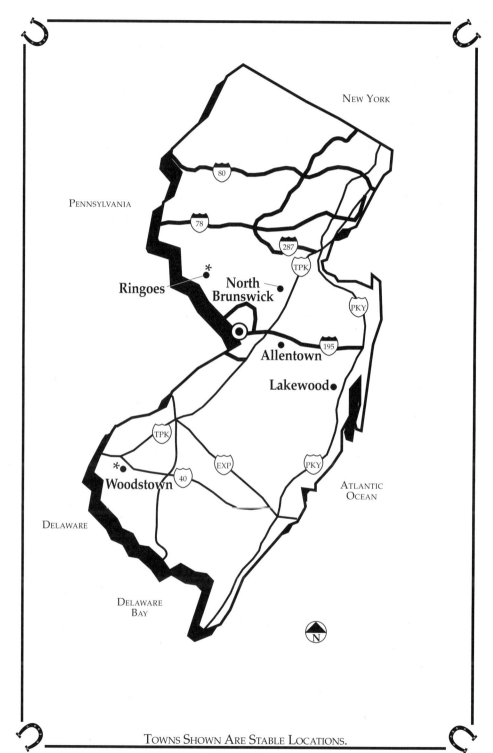

Towns Shown Are Stable Locations.

* onsite accommodations

NEW JERSEY Page 149

ALL OF OUR STABLES REQUIRE CURRENT NEG. COGGINS, CURRENT HEALTH PAPERS, & OWNERSHIP PAPERS.

ALLENTOWN
O-NO Acres Phone: 609-259-2334
Sandra Baggitt
45 Imlaystown Road [08501] **Directions:** Exit 11 off of I-95. Call for directions. **Facilities:** 10 indoor stalls, 6 paddocks, feed/hay & trailer parking available. Call for reservations. **Rates:** $15 per night. **Accommodations:** Motels 10 miles from stable.

LAKEWOOD
Lakewood Boarding Stables Phone: 908-367-6222
Bill M. Eak
436 Cross Street [08701] **Directions:** Exit 21 off of 195. Take 527 south to 528 east. Take right on Cross St. Stable is approx. 1 mile on right. **Facilities:** 85 indoor 10' x 12' stalls, 2-1/4 acre paddocks, 2 acres of pasture/turnout, giant indoor arena, quarantine area, feed/hay & trailer parking available. Stallions can be accommodated. 600 acres of riding trails. Riding lessons given. Reservations required. Tack shop on premises. **Rates:** $20 per night. **Accommodations:** Ramada Inn in Lakewood, 5 minutes from stable.

RINGOES
✶ **Cross Creek Farm** Phone: 908-806-3248
Loretta McCay
45 Dutch Lane [08551] **Directions:** 5 minutes off 202; easy access to Rts. 95, 78, and 287. Call for exact directions. **Facilities:** 10 indoor 10' x 10' stalls, 120' x 250' ring, 3 acres of pasture, wash stall, tack room, feed/hay available, trailer parking. Trail riding. Camping with bathroom accommodations, 3 campsites for trailers, water hookups. Petting zoo. **Rates:** $20 per night, $75 per week. **Accommodations:** Cottage on premises. Ramada Inn in Flemington, 3 miles from stable; other motels and B&B within 5 miles.

NORTH BRUNSWICK
Farrington Farms Phone: 732-821-9844
Gary Ippoliti E-mail: gary@farringtonfarms.cjb.net
Web: farringtonfarms.cjb.net
28 Davidson Mill Road [08902] **Directions:** Exit 8A off of NJ Turnpike. Call for directions. **Facilities:** 60 indoor stalls, round pen, 2 outdoor arenas, 100' x 250' & 100' x 100', 7 outdoor paddocks, indoor ring, jump course. Training of horses & riders for hunter/jumper, hunt seat, & English & Western pleasure. Breeds thoroughbreds. Call for reservations. **Rates:** $25 per night. **Accommodations:** Motels 2 miles from stable.

✶ ✶ ✶

PAGE 150 NEW JERSEY

ALL OF OUR STABLES REQUIRE CURRENT NEG. COGGINS, CURRENT HEALTH PAPERS, & OWNERSHIP PAPERS.

WOODSTOWN

✱ <u>**Victorian Rose Farm Bed & Breakfast**</u>   **Phone: 856-769-4600**
947 Rt 40 [08098]   Directions: At mile marker 8 1/2 on Rt 40 just east of Delaware Memorial Bridge. Or, Exit #1 off of the New Jersey Turnpike.
Facilities: Turnout pasturage only, no stalls, feed/hay available, trailer parking.
Rates: For B&B guests: $5 per night, $10 per night with feed. Overnight stabling only: $20 per night. **Accommodations:** 4-bedroom Victorian B&B on site, 2 rooms w/private bath, 2 rooms share a bath.

NOTES AND REMINDERS

Page 152 **NEW MEXICO**

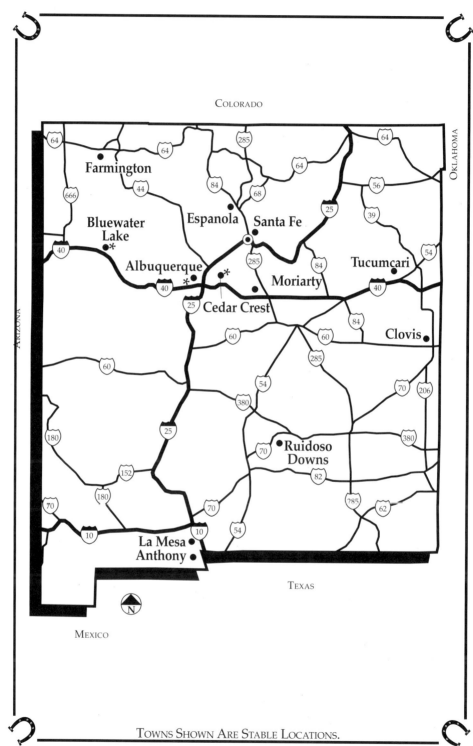

Towns Shown Are Stable Locations.
* onsite accommodations

NEW MEXICO PAGE 153

<u>ALL OF OUR STABLES REQUIRE CURRENT NEG. COGGINS, CURRENT HEALTH PAPERS, & OWNERSHIP PAPERS.</u>

ALBUQUERQUE
✱ <u>Blakley's Spur Stable</u>   Phone: 800-305-1851 or 505-877-1851
Clarinda Blakley   E-mail: spurstable@aol.com
2029 Lakeview SW [87105-6103] **Directions:** South of I-40 on I-25, Exit 220; go west 2.5 miles to Isleta Blvd. (3rd light); go south 1 mile to Lakeview; go west 1/4 mile. Stable is on right. **Facilities:** 7 indoor stalls, 6 outdoor stalls, & 15 stalls with runs, horse walker, riding trails, feed/hay & trailer parking available. Vet on call. Camper hook-up. Owners on premises. Reservations required! **Rates:** $15 per night; $70 per week. **Accommodations:** Guest house on premises. Motel 6 in Albuquerque, 5 miles from stable.

<u>Heartlane Farms, Inc.</u>   Phone: 505-345-7072
Julie & Bob Luzicka
6730 Rio Grande Blvd. NW [87107] **Directions:** From I-40 West: Take Rio Grande Exit; go north (right) for 4 miles. Barn is on east side, past fire station with sign out front. **Facilities:** 12 indoor stalls with shavings & outdoor runs, 8 indoor stalls, 6 outdoor stalls, 24' x 48' turnouts, large shaded arena, wash rack, crosstie and grooming area, feed/hay & trailer parking available. **Rates:** $25 per night for indoor stalls; $15 per night for outdoor stalls. $100 per week. Owner on premises. **Accommodations:** Sheraton Old Town & Best Western Inn at Rio Rancho, 5 miles away.

<u>Town-n-Country Feed & Stables</u>   Phone: 505-296-6711
Judy Davis
15600 Central SE [87123] **Directions:** Exit 170 off of I-40E. Call for directions. **Facilities:** 10 open pens. Complete feed & tack store on premises. Open 7 days a week. **Rates:** $10 per night.

ANTHONY
<u>Spur-C Ranch</u>   Phone: 505-874-3603
Boyd & Nancy Carson
Rt. 1, Box 617-L [88021] **Directions:** Exit 8 off of I-10. Go west about 3/4 mile to stop light; turn right on Doniphan; go 1/2 mile to stop light, turn left on Borderland Rd.; go 3.2 miles; at the third stop sign the road T's; sign for ranch is to the right & across the road; follow the signs on the dirt road to ranch gate that is 1.1 mile from pavement. **Facilities:** Call ahead for availability. Stalls are outdoor 50' x 50' with shades, 130' x 330' turnout, feed/hay & trailer parking. **Rates:** $10 per night. **Accommodations:** Motels in El Paso, 10 miles away.

ALL OF OUR STABLES REQUIRE CURRENT NEG. COGGINS, CURRENT HEALTH PAPERS, & OWNERSHIP PAPERS.

BLUEWATER LAKE
✱ **Bluewater Lake Lodge, RV and Horse Motel**   Phone: 505-290-1400
Patricia M. Jacoby   Fax: 505-285-4312
608 Hwy 412 [87045] **Directions:** I-40 to exit 63 Hwy 412(Prewitt), go South 7 miles to Park on right. **Facilities:** 5 holding pens with shelter, feed, trails. 5 full hookups with water & electric. RV space with stall and wash station, Zuni Mountains and Blue water Lake riding (free range). Guide available if desired. **Accommodations:** Cabins available in the village for rent nearby, while lake lodge is under construction.

CEDAR CREST/ALBUQUERQUE
✱ **Cedar Crest Country Cottage & Stables**   Phone: 505-281-5197
Donald & Annette Romeros   Cell: 505-980-7429
E-mail: dnacccc@aol.com   Web: www.cedarcrestcottage.com
47 Snowline Road, P.O. Box 621 [87008] **Directions:** Just 15 minutes from Albuquerque. East I-40 to Exit 175. Hwy 14 to Cedar Crest. Go north 4 miles (4 lane hwy). Turn left just past mile marker 4, Snowline Road. Go 1/2 mile past stop sign. White pipe fence. Check in - 1st driveway on left. #39, The Romeros. **Facilities:** 4 outdoor and 2 indoor covered stalls, round pen, 200' x 200' turnout paddock. Hay & trailer parking available. Adjacent to Cibola National Forest & trails. **Rates:** $15 per horse. From $75 Double occupancy in cottage - Call for reservations. **Accommodations:** 3 bedroom 2 bath cottage. Fully self sufficient kitchen.

CLOVIS
Amigo Del Caballo Horse Motel   Phone: 505-742-1033
Bob Meisenheimer
945 Curry Road-E [88101] **Directions:** 3 miles west of the Texas state line on Hwy 60-70 & 84; 4 miles east of Curry County Fairgrounds. Turn north at sign and go north 1/3 mile. **Facilities:** All pipe and concrete stalls with a "Shoofly" fly control system in all 7 stalls. Each 13' x 13' stall opens into an additional 13' x 27' run. There are two 50' x 50' six foot high pens, two 50' x 120' pens 4' high. More pens 50' x 300' and larger. A 335' x 205' lighted roping arena as well. Feed/hay and trailer parking available with 2 RV water and electrical hookups. Owner lives on site. **Rates:** $15 per night. $10 per day weekly rate. **Accommodations:** Bishops Inn, Clovis Inn, Comfort Inn, Days Inn, Holiday Inn, Kings Inn, LaVista Inn, Motel 6, Sands Motel on Hwy 60-70-84, 3-4 miles from stable.

Curry County Fairgrounds   Phone: 505-763-6502
Gary Hillis, Caretaker   or: 505-762-8827
600 South Norris [88101] **Directions:** At the junction of US 60, US 70, & US 84. Call for directions. **Facilities:** 12 outdoor 20' x 20' pens with no shelter, 18 indoor stalls, rodeo arena. 24-hr security. Rodeos, bull-riding, barrel racing, roping, circus, & Special Olympics are held at the fairgrounds. **Rates:** Donation requested for outside pens; $5 for inside stalls per night. **Accommodations:** Days Inn 1/2 mile from fairgrounds.

NEW MEXICO   PAGE 155

<u>ALL OF OUR STABLES REQUIRE CURRENT NEG. COGGINS, CURRENT HEALTH PAPERS, & OWNERSHIP PAPERS.</u>

ESPANOLA
<u>Roy-El Morgan Farm</u>   Phone: 505-753-3696
Elberta Honstein   Web: www.roy-elmorgans.com
E-mail: elberta@roy-elmorgans.com
1302 N. McCurdy Road [87532] **Directions:** Call for directions. **Facilities:** 28 12'x12' indoor box stalls, 5 large outdoor runs, feed/hay and trailer parking available. Show and training, state of the art stallion station; we stand 3 stallions and also stand outside stallions and ship semen. **Rates:** $15 per night. $10 per day- weekly rate. **Accommodations:** Motel 8, Best Western, Holiday & Comfort Inn no more than 6 miles away.

FARMINGTON
<u>Las Brisas</u>   Phone: 505-564-8948
Donnie & Sherry Pigford   505-327-1855
2446 LaPlata Hwy [87401] **Directions:** Hwy 64 to Shiprock. Right on Hwy 170 (La Plata Hwy) 1 mile N on right side of Hwy. **Facilities:** 7 indoor stalls with 16' runs and 9 outdoor stalls with covers and runs. Feed available. Trailer parking. Hot walker, wash rack, arena, round pen. 1 camper hook-up with water and electric $15 per night. **Rates:** $10 per night for outdoor stall and $15 for indoor. **Accommodations:** Several motels 1-5 miles away.

<u>McGee Park Fairgrounds</u>   Phone: 505-325-5415
Jim Parnell, Director
41 Road 5568 [87401] **Directions:** Fairgrounds is visible from Hwy 64 behind race track. Call for directions. **Facilities:** 100 outside covered 12' x 12' stalls, 100' x 290' indoor coliseum, 100' x 200' outdoor arena. Concerts, rodeos, roping, horse shows, etc. held at fairgrounds. Open 24 hours. Office hours: 8-4:30 M thru F. **Rates:** $5 per night. **Accommodations:** Motel 6.5 miles away.

LA MESA
<u>Armstrong Equine Services</u>   Phone: 505-233-2208
Joe B. Armstrong
175 Esslinger Rd [88044] **Directions:** Exit 155, Vado, off of I-10. Call for further directions. **Facilities:** 53 indoor stalls, wash racks, arenas, walker, paddock, electric hook-up for self-contained campers. Feed/hay & trailer parking available. Horse training and stud service available. **Rates:** $20 per night. **Accommodations:** Motels in Las Cruces & in Anthony, Texas.

NEW MEXICO

<u>ALL OF OUR STABLES REQUIRE CURRENT NEG. COGGINS, CURRENT HEALTH PAPERS, & OWNERSHIP PAPERS.</u>

MORIARTY
<u>Rockin' Horse Ranch</u>   Phone: 505-832-6619
Lonnie & Patty Wright   Fax: 505-832-6619
28 Western Rd [87035] **Directions:** Take I-40 to exit 196(Hwy 41). Turn north on Hwy 41 go 4.5 miles to Western Rd. Turn right, first driveway on right. **Facilities:** 32 indoor 12x12 stalls. 10 runs, semi covered. Feed and hay available, trailer parking available. Indoor and outdoor arena for turnout. **Rates:** $20 per night. **Accommodations:** Holiday Inn Express, Super 8, and Motel 6 in Moriarty.

RUIDOSO DOWNS
<u>Tull Stansell Estate</u>   Phone: 505-378-4503
Cheryl & June McCutcheon   Barn Manager: 505-378-8188
P.O. Box 57 [88346] **Directions:** Located off of Hwy 70. Call for directions. **Facilities:** 9 indoor stalls, 5 stalls with runs, 10 paddocks, hot walker. Boards & sells llamas. Reservations requested. **Rates:** $5 per night; $21 per week. **Accommodations:** RV campground nearby; Inn at Pine Springs, 2 miles away.

SANTA FE
<u>Northern New Mexico Horsemen's Association</u>   Phone: 505-471-6654
Web: horse-talk.com/nnmha   E-mail: NNMHA@horse-talk.com
PO Box 4124 [87502] **Directions:** NNMHA Arena-Rodeo-Fair Grounds on Rodeo Dr. **Facilities:** 150 outdoor stall, must provide own care. **Rates:** Call for information. **Accommodations:** Many hotels in area.

TUCUMCARI
<u>Western Drive Stables</u>   Phone: 505-461-0274
Jim & Marlene Haller
PO Box 1072 [88401] **Directions:** North of I-40 exit 331/Camino del Coronado. Close to Route 66. Call for directions. **Facilities:** 20- 12' x 18' covered outdoor and boxed stalls, stallion stalls available. Holding pens, paddocks, exercise area, space for trailers, electric hook-up only. All weather driveway & feed & bedding available. parking with electric for campers and trailers, all weather driveway,Vet close by. Owners on premises. Reservations encouraged, but not necessary. **Rates:** Please call. **Accommodations:** Close to motels and restaurants.

NOTES AND REMINDERS

NEW YORK

TOWNS SHOWN ARE STABLE LOCATIONS.

* onsite accommodations

NEW YORK PAGE 159

ALL OF OUR STABLES REQUIRE CURRENT NEG. COGGINS, CURRENT HEALTH PAPERS, & OWNERSHIP PAPERS.

ADAMS CENTER
Royal Stables   Phone: 315-583-6429
Mary Ramsey
Green Settlement Road [13606] **Directions:** 1/2 mile off of I-81 at Exit 42. Call for directions. **Facilities:** 27 indoor stalls, 2 acres of pasture/turnout, 60' x 110' indoor arena, 100' x 200' outdoor arena, round pen, 4 paddocks, feed/hay & trailer parking on premises. Beautiful, scenic riding trails. Full care boarding facility that also offers training of horses & riders in English, Western, jumping, etc. at all levels. Must have rabies certificate. Call for reservation. **Rates:** $15 per night. **Accommodations:** Days Inn & Ramada Inn 6 miles from stable.

BIG FLATS
Gale's Equine Facility   Phone: 607-796-9821
Gale Wolfe
219 Sing Sing Road [14845] **Directions:** Call for directions. **Facilities:** 5 indoor stalls, two 100' x 200' turnout areas, indoor arena, feed/hay & trailer parking available 24-hr advance notice preferred. **Rates:** $15 per night. **Accommodations:** Howard Johnson's & Holiday Inn in Horseheads, 15 minutes from stable.

BINGHAMTON/PORT CRANE
Pleasant Hill Stable   Phone: 607-648-4979
Rose Illsley
648 Pleasant Hill Rd [13833] **Directions:** 1.25 miles off I-88. Exit 4 (9 miles North East of Binghamton. Call for directions. **Facilities:** 80 10' x 12' box stalls, indoor and outdoor riding arena. Feed/hay available, trailer parking available. Complete tack shop over indoor riding ring, miles of trails for riding. Thoroughbred and Belgian Stallion. **Rates:** $15 per night, $75 per week. **Accommodations:** Days Inn and Front Street Motel 6, 6 miles from stable in Binghamton, NY.

BUFFALO
Buffalo Equestrian Center   Phone: 716-877-9295
Libby McNabb   E-mail: sbeat@aol.com
950 Amherst St [14216] **Directions:** 90 to 33 West to 198 West to 1st Delaware Ave. exit right at stop sign, left at light. Go 1/4 mile to Great Arrow turn left, we are located on left. **Facilities:** 110, 10'x10' wood stalls. Minimum trailer parking. Horse shows and lesson program. **Rates:** $35 per night. **Accommodations:** Holiday Inn (716-886-2121).

ALL OF OUR STABLES REQUIRE CURRENT NEG. COGGINS, CURRENT HEALTH PAPERS, & OWNERSHIP PAPERS.

CHERRY CREEK
✽ **Foxe Farmhouse Bed & Barn** Phone & Fax: 716-962-3412
Carol Lorenc Toll Free: 877-468-5523
Web: www.foxefarmhouse.com E-mail: lorencfoxefarm@madbbs.com
1880 Thornton Rd. [14723] **Directions:** 20 minutes from I-90 and I-86, Call for directions. **Facilities:** 5 12x12 indoor stalls, 3 8x12 stalls attached to barn, feed/hay available, trailer parking, a one acre paddock with run-in sheds, 1-3 acre Pasture with run-in., saw dust and bedding, turnout, 225x100 sand arena, scenic country roads for riding and carriage driving. Near Amish country and Cockaigne Ski Center and Chautagua Institution. Please Call Ahead! **Rates:** $15 per night, $50 per week, $300 per month, includes grain and hay. **Accommodations:** Special Lodging/Stabling Rates! Foxe Farmhouse provides individual rooms or exclusive use of our entire house (sleeps 10) with full kitchen, living room, and porches.

HAMMONDSPORT
✽ **Donameer Farm** 607-569-2115
Cynthia Harrison, Neal Esposito E-mail: donameer@empacc.net
7417 Smallige Road [14840] **Directions:** I-86 to exit 40 (Savona) 415N to Robie Rd. Follow 2 miles, take left at fork onto Velie Rd. Velie Rd. Turns into Smallige Rd. Farm on left. **Facilities:** 10- 12x12 covered stalls. Hay, trailer parking available. 10 fenced in acres, 8- 100' X 100' foot paddocks, 7 grassy 100'x110' paddocks, 12-30 amp/water hookups, dump station, miles of trails. 10 miles from Sugar Hill State Park; 50 miles of trails. Well-behaved dogs are welcome. Negative Coggins. **Rates:** stalls-$15/horse/night, paddocks-$5/night/horse, hookups-$25/night **Accommodations:** 1 furnished bedroom cottage, $60/weeknight, $65/weekend night. Days Inn/Bath and Vinehurst Motel each within 5 miles.

LAKE LUZERNE
Bennett's Riding Stable Phone: 518-696-4444
Lawrence & Bonnie Bennett
RR 2, Box 208 Gage Hill Road [12846] **Directions:** From I-87: Take Rt. 9N south towards Lake Luzerne for 5 miles. Stable is on the left-hand side. **Facilities:** 4 indoor stalls, 5 acres of pasture/turnout, sweet feed & hay available, trailer parking on premises, easy access to beautiful state riding trails plus manyother trails within short driving distance. Portable round pen also available. Reservation required. Guided horseback riding on their horses from 1 hour to all day. **Rates:** $15 per night; $100 per week. **Accommodations:** Kastner's Motel & Pine Point Motel & Cottages, & Nancy Lee Motel all within walking distance in Lake Luzerne.

LEROY
Jensen Stable Phone: 585-223-9508
Abigail Jensen
7077 West Main Street [14482] **Directions:** I-90 to LeRoy Exit, south on Rte. 19 into LeRoy, turn right at light onto Rte. 5, stable 1 mile on right. **Facilities:** Box stalls, pastures, paddocks, indoor arena available. Feed, hay and trailer parking available. Certified instructor. Late arrivals welcome. **Rates:** $25 per night. **Accommodations:** Motels and Hotels close by.

NEW YORK Page 161

ALL OF OUR STABLES REQUIRE CURRENT NEG. COGGINS, CURRENT HEALTH PAPERS, & OWNERSHIP PAPERS.

LITTLE FALLS
Diamond Hill Phone: 315-429-3527
Adam (Mike) Miller Barn Manager: 315-429-3514
RD 1, Dairy Hill Road [13365] Directions: Exit 29A on NY State Thruway (Little Falls) to Rt. 169; make right at flashing light; make left at Rt. 167; travel thru Dolgeville; bear left on State St.; go 3.4 mi. to stop sign; cross intersection; go 1.9 mi. Diamond Hill on left. **Facilities:** 15 indoor 12' x 12' box stalls, 7 paddocks of various sizes that can accommodate 20 horses, auto waterers, wash bay, jumping ring, sand warm-up ring, indoor arena, beautiful riding trails on property, tack shop on premises. To meet your needs, reservations are necessary. "We appreciate early arrivals." In the case of delays, please call ahead. This equestrian center has been host to dressage clinics, charitable events, shows and tack & horse auctions. **Rates:** $20 per night; $18 per day weekly rate. **Accommodations:** Best Western in Little Falls, 8 miles away, Adriana's B & B in Dolgeville, 5 miles away, Herkimer Campgrounds 14 miles from stable.

MONTGOMERY
Highland Farm Phone: 845-361-2204
Yvonne Turchiarelli Fax: 845-361-2025
E-mail: higlandfarm@fcc.net Web: www.highlandfarmny.com
2101 Rt. 17K [12549] Directions: From I-84 W: Take Exit 5, right turn onto Rt. 208 N; go 1 mile to light at intersection of 208 & 17K; turn left onto 17K & go 5.6 miles to farm on left. **Facilities:** 30 indoor stalls, 8 outdoor stalls, several varied sizes of pasture/turnout, paddocks, feed/hay & trailer parking available. This farm has a clean barn & experienced help. Vet nearby. **Rates:** $30 per night. **Accommodations:** Comfort Inn in Newburgh, 12 miles; Harvest Inn in Pine Bush, 8 miles; Holiday Inn, 8 miles from stable.

NEW PALTZ
Million Dollar Farm Ltd. Phone: 845-255-5877
Frank Heyer
300 Springtown Road [12561] Directions: NY Thruway (I-87) Exit 18: Left on 299; go thru New Paltz for 1.8 miles over bridge; take first right on Springtown Rd.; go 3.2 miles; farm is on right. **Facilities:** 6 indoor box stalls, 23 indoor straight 5' x 10' stalls, 125' x 220' pasture/turnout, feed/hay & trailer parking on premises. "We have the best trails in the county." (NYC aqueduct, Mohonk Trails, Railtrail, Minnewaska State Park, plus trails on the farm). **Rates:** $10 for straight stall; $15 for box stall. **Accommodations:** Super 8, Day Stop Inn, Motel 87, & EconoLodge all in New Paltz, 5 miles from stable.

NEW YORK

ALL OF OUR STABLES REQUIRE CURRENT NEG. COGGINS, CURRENT HEALTH PAPERS, & OWNERSHIP PAPERS.

PLATTSBURGH
Cedar Knoll Farm   **Phone: 518-561-6003**
Susan Castine
246 Spellman Road [12901]   **Directions:** Exit 40 off I-87, 1/4 mile west; 30 miles south of Canadian border. **Facilities:** 4 indoor 12' x 12' box stalls, 40-acre pasture, 200' x 300' paddock, feed/hay, trailer parking available, call for reservations. **Rates:** $25 per night, $100 per week. **Accommodations:** Restaurant and motel within 1/4 mile.

PORT BYRON
Snug Horse Haven   **Phone: 315-776-8243**
Jeri Marshall
RD #1 Maiden Lane Road [13140]   **Directions:** Off of I-90 at Exit 40. Call for directions. **Facilities:** 4 indoor stalls, two 75' x 100' paddocks, large round pen, 120' x 200' outdoor ring, full jump course, feed/hay & trailer parking on premises, horse training available at this small comfortable facility. Call for reservation. **Rates:** $15 per night. **Accommodations:** Port Forty Motel & Best Western 5 miles away.

PORT CRANE/BINGHAMTON
Pleasant Hill Stable   **Phone: 607-648-4979**
Rose Illsley
648 Pleasant Hill Rd [13833] **Directions:** 1.25 miles off I-88. Exit 4 (9 miles North East of Binghamton. Call for directions. **Facilities:** 34 -10' x 12' box stalls, indoor and outdoor riding arena. Feed/hay available, trailer parking available. Complete tack shop over indoor riding ring, miles of trails for riding, extra horses for riding. Thoroughbred and Belgian Stallion. **Rates:** $15 per night, $75 per week. **Accommodations:** Days Inn and Front Street Motel 6, 6 miles from stable in Binghamton, NY.

SARANAC LAKE
Sentinel View Stables   **Phone: 518-891-3008**
Carol Shante
Harriettstown Road [12983]   **Directions:** Located on Rt. 86. Call for directions. **Facilities:** 5 indoor stalls, 3 holding paddocks, riding trails on top of the Adirondack Mountains, trailer parking available. English and Western riding lessons offered. Call for reservation. **Rates:** $10 per night; ask for weekly rate. **Accommodations:** Motels 3 miles from stable.

NEW YORK PAGE 163

ALL OF OUR STABLES REQUIRE CURRENT NEG. COGGINS, CURRENT HEALTH PAPERS, & OWNERSHIP PAPERS.

STUYVESANT
* **(Westwind Farm)Roway Garden House** **Business: 518-758-6855**
Vivian A. Cook **Residence: 518-758-2083**
545 Rt 26A [12173] Directions: Exit 12 off I-90, take rte 9 South through Kinderhook Village, from village green traffic light, 1.5 miles to rt. 26A, bear right at Y, half mile on right. **Facilities:** 2 10'x12' indoor stalls with steel meshed doors and mats, hay and trailer parking available, In 7 out available, 1 acre paddock, round pen. Roway Garden House at Roway Farm is a 3-room B&B with private baths in the heart of the Hudson Valley, close to Saratoga and the Chatham Hunt Club. **Rates:** $15 per night, $75 per week. **Accommodations:** B&B on site.

WARWICK
Meadowood Farm, Inc. **Phone: 845-986-7387**
Cindy Van der Plaat
82 Belcher Road [10990] Directions: 20 minutes off of Rt. 17. Call for directions. **Facilities:** 14 indoor box stalls, two 2-acre pastures, two 1/2-acre pastures, 60' x 160' indoor arena, dressage arena, jumping field, round pen, 5 outdoor paddocks, heated wash stall, trailer parking on premises. Training of horses & riders in event riding and hunter/jumper. Call for reservation. **Rates:** $15 per night; ask for weekly rate. **Accommodations:** Motels within 7 miles of stable.

PAGE 164   **NORTH CAROLINA**

TOWNS SHOWN ARE STABLE LOCATIONS.

* onsite accommodations

NORTH CAROLINA PAGE 165

ALL OF OUR STABLES REQUIRE CURRENT NEG. COGGINS, CURRENT HEALTH PAPERS, & OWNERSHIP PAPERS.

BREVARD
✷ Free Rein Theraputic Riding Center   Phone: 828-883-3375
Nancy Searles
Rt. 1, Box 12-A [28712]   **Directions:** From I-26: Take Asheville Exit 9 (Airport exit) then Rt. 280W to center of Brevard (Broad & Main approx. 18.5 miles from airport); take 276 South for 7.1 miles then left onto See Off Mountain Rd.; go 1.7 miles to Las Praderas sign. **Facilities:** 33 indoor box stalls, feed/hay & trailer parking available, no tractor-trailers, please, at least 24-hr notice. **Rates:** $20 per night; $100 per week. **Accommodations:** Beautiful lodging available for a 2-night minimum stay. Also Imperial Motor Lodge in Brevard, 10 miles from stable. Call for free brochure.

BRYSON CITY
Double Eagle Farm   Phone: 828-488-9787
Greg & Karen Crisp
50 Sawmill Creek Road [28713]   **Directions:** 1 mile off Hwy 74 (4-lane), easy access. Call for directions. **Facilities:** 14 indoor 12' x 12' stalls each with window, skylight, ceiling fan; several 1/2-acre turnouts; feed/hay & trailer parking available. Close to many trails in Great Smoky Mountains National Park, Nantahala National Forest, Pisgah National Forest. Trail information & maps available. Call for reservations. **Rates:** $20 per night; weekly rate negotiable. **Accommodations:** Many motels nearby. Will mail list.

CASTLE HAYNE
Castle Stables, Inc.   Phone: 910-675-1113
Debi Mastrangelo
1513 Sidbury Road [28429]   **Directions:** 2-3 miles from Exit 414 off of I-40. Call for directions. **Facilities:** 10 indoor 12' x 12' & 12' x 14' stalls, 11 paddocks & pastures, 2 indoor wash racks with hot water, 2 riding rings, 1 with jumps and lighting. 13 miles from beach. No stallions without prior discussion. 24-hr advance notice. **Rates:** $15 including feed; weekly rate negotiable. **Accommodations:** Fairfield Inn & Days Inn 5 miles from stable.

DALLAS
Rose Hill Farm & Equestrian Training Center   Phone: 704-922-0866
Anna Lyman
702 Ike Lynch Road [28034]   **Directions:** From I-85, take 321N to Hardin Rd. exit; turn left & go straight 1.5-2 miles. Hardin Rd. turns into Ike Lynch Rd. **Facilities:** 5 indoor stalls, pasture/turnout areas, feed/hay & trailer parking available. Historic areas and Dallas Horse Park 5 minutes away, as much notice as possible. **Rates:** $15 including feed. **Accommodations:** Hampton Inn & Holiday Inn within 10 minutes.

ALL OF OUR STABLES REQUIRE CURRENT NEG. COGGINS, CURRENT HEALTH PAPERS, & OWNERSHIP PAPERS.

DUNN
Aysgarth Stables   Phone: 910-892-4030
Faith Bradshaw
129 Bumpas Creek Access [28334] **Directions:** I-95 exit 71. 4 miles off 95. Call for directions. **Facilities:** 6 indoor stalls, turnout, pasture and shelter on 10 acres. Hay/feed available. Nice trails within short trailering distance. Trailer and semi parking available. Reservations required. **Rates:** $20 per night. **Accommodations:** Fully equipped apartment sleeps 5.

GODWIN/DUNN
Stewart Horse Farm   Phone 910-980-1967
Ray Stewart Dunn Rd.
9220 [28344] **Directions:** Exit 65 off 1-95 west on 82 Hwy. Left turn at flashing light. Farm on left 1 mile from light. Call for directions. **Facilities:** 6 10x12 inside stalls, round pen, 3 paddocks. Feed/hay available. Black reg. Q.H. stud. Great trail riding. **Rates:** $25 per night. **Accommodations:** Many motels in Dunn.

LAUREL SPRINGS
✱ **Laurel Springs Farm**   Phone: 866-730-6410/336-982-3320
Nancy E. Smith   Web: www.laurelspringsfarm.info
E-mail: nancy20819@aol.com
455 Hiram Bare Rd [28644] **Directions:** From I77, N on 421, 421 Busns to Hwy 18 N to Blue Ridge Pkwy, south on BRP 1.9 miles, right on Hiram Bare, 1/2 mile on right, red roof on barn. **Facilities:** 6 indoor, 12'x12', 1 stud stall. 80'x80' wood board paddock, round pen with all weather footing. Feed/hay available, trailer parking available with RV hook ups and disposal station. Farm is open May 1 to October 20. Closed in winter. **Rates:** $10 per night self-care, bedding & feed extra. **Accommodations:** Romantic B&B on premises. There are several motels, B&B's and rental cabins nearby. Golf resort 15 min away.

LENOIR
✱ **Moore's Horseplay Ranch**   Phone: 828-757-9114
Everette & Kathy Moore
1154 Woodrow Place [28645] **Directions:** I-40 to 64/90. Less than 1 mile off Hwy 64/90. In the foothills of Blue Ridge Parkway. Call for directions. **Facilities:** New stall barn, 4 indoor 10' x 10' stalls, 6 outdoor 3' x 8' tie-up stalls, corrals, feed/hay & trailer parking available. Room for oversized trailers. Ranch offers riding lessons & miles of riding trails in a beautiful vacation area offering camping, fishing, hiking, etc., unlimited area for horse camping for portable stalls, cross ties, & picket lines, 9 indoor stalls in camping area. Western Town & tack shops nearby. Beautiful setting for horse vacations. Home of black & white national spotted saddle horse stallion: "Sepper Go Devil" (Go Boy). Must have reservations. **Rates:** $10-$20 per night. **Accommodations:** Log cabins for rent on premises plus large campsites. Also Holiday Inn & Days Inn in Lenoir, 7 miles from ranch.

NORTH CAROLINA Page 167

ALL OF OUR STABLES REQUIRE CURRENT NEG. COGGINS, CURRENT HEALTH PAPERS, & OWNERSHIP PAPERS.

MEBANE
Brindabella Farms LLC Phone: 919-304-3473
Enid Kafer Pager: 1-800-458-2397 Fax: 919-304-3424
E-mail: enidkafer@mindspring.com Web: www.brindabellahorses.com
3720 Mebane Oaks Road [27302] **Directions:** I-85/40 to exit 154, Go South for 4.2 miles, entrance to the farm is on the left. **Facilities:** 20-12x12 stalls in stable, 12x24 covered corrals, trailer parking, please bring hay/feed, Breed/Train/Show Sport Horses: Trakehner, Arab, and TB. Negative Coggins and Ownership. Please confirm reservation **Rates:** $35-45 per night, Visa, MC accepted. **Accommodations:** Hampton Inn, Holiday Inn Express nearby.

RALEIGH
Triton Stables Phone: 919-847-4123
Ellen Welles Evenings: 919-847-5446
9901 Macon Road [27613]
Directions: From I-85 S, take Exit 86 for Creedmore/Buttoner; take Hwy 50 towards Raleigh for 12 miles; at stop light turn left on Norwood; go 1.75 miles and take left on Macon, stable is 1 mile on left. **Facilities:** 5 indoor stalls, no pasture/turnout. Call for reservation. **Rates:** $15 per night w/o feed.

RUFFIN
Chestnut Hill Stables & Riding School Phone: 336-939-7126
Carole Moore
630 Mayfield Road [27326] **Directions:** From I-29 N, take Mayfield Road exit; turn left at end of ramp. From I-29 S take right at end of ramp. Stable is 3/4 mile on left. **Facilities:** 10 indoor stalls, pasture/turnout, feed/hay at additional charge, and trailer parking available. Reservation preferred but will help in an emergency with no notice. **Rates:** $10 per night. **Accommodations:** Holiday Inn and Comfort Inn 8 miles from stable.

TRINITY
✱ **Never Done Acres** Phone: 910-475-1914
Michael O'Neill & Pamela Merritt
3124 Finch Farm Road [27370] **Directions:** 4 miles south of Exit 106 (Finch Farm Road) off of I-85. Call for directions. **Facilities:** 3 indoor stalls, three 2-acre pastures, lighted riding ring, feed/hay, & trailer parking available. Vet & farrier on call. Camper hook-ups. Dog pen available. Call in advance. **Rates:** $15 for stall per night; $5 for pasture. **Accommodations:** Bed & Breakfast on premises, $20 per room & no smoking in house. Days Inn & Ramada Express in Thomasville, 6 miles away.

WAYNESVILLE
Dye-Na-Mite Show Barn Phone: 704-627-2666
Martha Dye, Owner, Nancy Campbell, Manager
300 Witch Way [28786] **Directions:** 2.5 miles from I-40 at Exit 24. Call for directions. **Facilities:** 10 indoor 12' x 12' stalls, 84' x 100' indoor arena, small pasture, feed/hay & trailer parking available. Reservation required. **Rates:** $15 per night; $75 per week. **Accommodations:** Several motels 6 miles away.

Page 168 **NORTH DAKOTA**

Towns Shown Are Stable Locations.

* onsite accommodations

NORTH DAKOTA Page 169

ALL OF OUR STABLES REQUIRE CURRENT NEG. COGGINS, CURRENT HEALTH PAPERS, & OWNERSHIP PAPERS.

BISMARCK
Selland Indoor Arena Phone: 701-255-1420
Lee Selland
6600 N.E. 26th Street [58503] **Directions:** Located off of I-94. Call for directions. **Facilities:** 68 indoor stalls, outside pens, indoor arena, 2 outdoor arenas, feed/hay & trailer parking available. Open 24 hours daily. Call ahead for reservation. **Rates:** $10 per night; weekly rate available. **Accommodations:** Embassy Suites & Ramada Inn 5 minutes from stable.

DEVILS LAKE
Rush Valley Farms Phone: 701-662-4386
Kevin Frith
Rt. 3, Box 340 [58301] **Directions:** Located off of Old Hwy 2, two miles outside of town. Call for directions. **Facilities:** 12 indoor stalls, pasture/turnout area, feed/hay & trailer parking available. Call for reservation. **Rates:** $10 per night. **Accommodations:** Many chain motels 2 miles from stable.

DICKINSON
North Forty Stables Phone: 701-227-1915/701-483-4886
Prairie Ventures LLC
Sylvia Hartford E-mail: Hartford@goesp.com
11042-32nd "R" St. SW [58601] **Directions:** I-94 take exit 61 on Hwy 22, north 4 miles, 1/2 mile on south side of road. **Facilities:** 23 stalls, 15 indoor, 8 outdoor runs with shelter. Outdoor riding arena. Feed/hay available, trailer parking available. **Rates:** $15 per night. **Accommodations:** Several motels 4 miles from Dickinson. Close to North Dakota Badlands and Maah Daah Hey Trail.

FARGO
Breezy Manor Equestrian Centre Phone: 701-280-0073
Buel Sonderland or: 701-282-9582
Web: www.breezymanor.org
6340 77 St. S [58104] **Directions:** 3 miles West of exit 60, off I-29, or 4 miles South of exit 346A off I-94. **Facilities:** 25 indoor 10'x12' stalls, feed/hay available, trailer parking, six single- acre paddocks. This facility is for sale and could serve as a boarding stable and serve as the Race Horse Industry between Winnepeg and Minneapolis/St. Paul. **Rates:** $15 per night. Please visit our website at www.breezymanor.org for further information. **Accommodations:** Fargo (4 miles).

NORTH DAKOTA

ALL OF OUR STABLES REQUIRE CURRENT NEG. COGGINS, CURRENT HEALTH PAPERS, & OWNERSHIP PAPERS.

FORT RANSOM
Fort Ransom State Park   Phone: 701-973-4331
John Kwapinski
5981 Walt Hjelle Parkway [58033-9712] **Directions:** Follow signs. 12 miles off Hwy 46, 34 miles from I-94 (Exit 292), 34 miles south of Valley City. **Facilities:** 16 - 16' x 16' board-fenced pens, room to set up portable corrals, water and trailer parking available, electric hook-ups nearby. 900-acre park in Sheyenne River Valley; wooded, hilly park with 5 miles of riding trails. **Rates:** $4 per night per horse. Vehicle entrance fee $4 per vehicle. **Accommodations:** Primitive camping near corral area $11 per night, $16 per night with electricity. Campsites with electricity are 1 mile from horse corral. Island Park Motel and Super 8 Motel in Lisbon, 21 miles away. Viking View Resort (cabins) 2 miles south of park.

GRAFTON
Thompson Stables   Phone: 701-352-2732
Charles Thompson   701-352-2410
527 West 15th Street [58237] **Directions:** Call for directions. **Facilities:** 2 indoor stalls, pasture/turnout area, feed/hay & trailer parking available. Call for reservation. **Rates:** $15 per night; weekly rate available. **Accommodations:** Super 8 two minutes from stable.

GRAND FORKS
Kuster's Wagon Wheel Stables   Phone: 701-772-6526
Myron & Joyce Kuster
Rt. 1, Box 226 [58201] **Directions:** Located 1 3/4 mile SW of Grand Forks city limits (I-29 & 32nd Ave. S). Call for directions. **Facilities:** 24 indoor stalls, 55' x 100' indoor riding arena, trailer parking. Call for reservation. **Rates:** $15 per night; negotiate longer stays, includes feed/hay. **Accommodations:** Many motels approx. 2 miles into town.

HANKINSON
Country Acres Veterinary Clinic   Phone: 701-252-7133
Dr. Barb Looysen
8279 37R Street S.E. [58401] **Directions:** Less than 1 mile from the Jamestown Exit off of I-94. Call for directions. **Facilities:** 4 indoor box stalls, 4 acres of fenced pasture/turnout, holding pen, feed/hay & trailer parking. Complete veterinary care available on premises. Call for reservation. **Rates:** $15 per night. **Accommodations:** Comfort Inn, Dakota Inn, & Super 8 all 1 mile away.

MEDORA
✶ **Little Missouri Horse Co.**   Phone: 701-623-4496
Wally Owen
P.O. Box 8 [58645] **Directions:** Off of I-94, 1 mile south of Medora, Theodore Roosevelt National Park and Sally Creek State Park. **Facilities:** 6 - 8 X 10 outdoor stalls with primitive camping next to Little Missouri River. Well-marked riding trails. **Rates:** $5 per horse. **Accommodations:** $8 per night camping, also 2 bedroom fully equipped cabin (sleeps 6) with private corrals. **Rates:** $95.75 per night plus $5.00 per horse. Motels within 6 miles.

NORTH DAKOTA   PAGE 171

ALL OF OUR STABLES REQUIRE CURRENT NEG. COGGINS, CURRENT HEALTH PAPERS, & OWNERSHIP PAPERS.

MINOT
Blue Hank Horse Stable   Phone: 701-839-3689
Al & Nancy Schall   E-mail:anschall@ndak.net
1615 27th St SE [58701] **Directions:** Call for directions. **Facilities:** 10-12 indoor and outdoor stalls. Feed/ hay available, trailer parking available. **Rates:** $10 per night. **Accommodations:** Many nearby.

Blue Hank Horse Stable

Phone: 701-839-3689
E-mail: anschall@ndak.net
1615 27th St SE, Minot, North Dakota (58701)

Page 172 **OHIO**

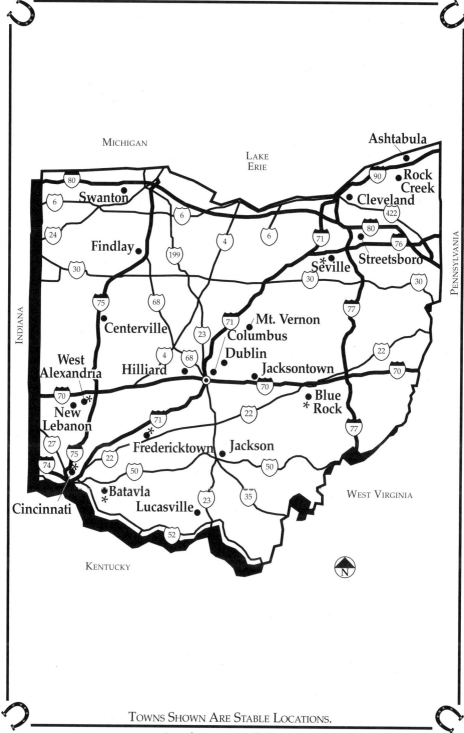

Towns Shown Are Stable Locations.
* onsite accommodations

OHIO PAGE 173

ALL OF OUR STABLES REQUIRE CURRENT NEG. COGGINS, CURRENT HEALTH PAPERS, & OWNERSHIP PAPERS.

ASHTABULA
Koch Show Horses Phone: 440-224-2097
Jim & Anne Koch
2560 Plymouth-Gageville [44004] **Directions:** I-90 to Kingsville exit (#235, Rt. 193 South) to second crossroad (Plymouth - Gageville) turn right. Farm is 1.8 miles on left. Large illuminated farm sign. **Facilities:** 35 stalls, all indoor. Size ranges from 10' X 10' to 12' X 12'. Outdoor turnout w/large run-in shed available. Feed, Hay and trailer parking available. Large circular drive and parking area. Seven individual turnouts w/split rail fencing, 1 acre each. Large indoor riding arena, Indoor Round Pen. Owner operated. Onsite supervision of horses at all times. The horse business is our only business. Forty years professional equine experience. **Rates:** $15 per day, $80 per week. **Accommodations:** Three motels 3 to 7 miles away.

BATAVIA
* **East Fork Stables & Lodge** Phone: 513-797-7433
George Wisbey
2215 Snyder Road [45103] **Directions:** Call for directions. **Facilities:** 32 indoor stalls, indoor arena, several fenced paddocks with run-in shelters, 50 miles of trails, feed/hay & trailer parking available. Horse rentals and trail rides. Reservation required and at least 24-hr notice. **Rates:** $20 per night. **Accommodations:** 1 cabin on premises available for rent. Holiday Inn & Red Roof Inn within 15 minutes.

BLUE ROCK
* **McNutt Farm II/Outdoorsman Lodge** Phone: 740-674-4555
Don R. & Patty L. McNutt
6120 Cutler Lake Road [43720] Farm with Lodging with continental breakfast, Overnight for humans and their critters. Advance reservation required with 50% deposit, balance in cash on arrival. No alcoholic beverages. **Directions:** 11 miles from I-70, 35 miles from I-77, & 55 miles from I-71. Exit 155 S -state rte. 60 S. **Facilities:** Secure stalls and trailer parking. Check-in 5-7 P.M.; check-out by 9 A.M. Overnight boarding is for lodge guests only. **Rates:** $40 per night per person, $10 per night per horse, pets allowed at $5.00. Ask about our vacationers rates for weekenders or by the week. **Accommodations:** Rooms at the Lodge type farm house, Log Cabin, Carriage House, the Cellar building at a vacation destination offering trail riding, hiking, fishing, etc. We pride ourselves on serving your needs and the needs of your traveling equine.

CENTERVILLE
Menker's Circle 6 Farm Phone: 513-885-3911
Robert Menker
11090 Yankee Street [45458] **Directions:** Call for directions. South suburban area of Dayton; 2.1 miles south of I-675, less than 1 mile from I-75; 35 miles north of Cincinnati. **Facilities:** 78 indoor 10' x 12' stalls, various sized turnouts, 60' x 130' indoor arena, outdoor arena, round pen, feed/hay, trailer parking, and self-contained campers welcome. Specializing in quarter horses. Call for reservations. **Rates:** $25 per night. **Accommodations:** Motels and shopping within 2 miles of stable.

PAGE 174 OHIO

ALL OF OUR STABLES REQUIRE CURRENT NEG. COGGINS, CURRENT HEALTH PAPERS, & OWNERSHIP PAPERS.

CINCINNATI
✴ **First Farm Inn** Phone: 859-586-0199
Jen Warner Web: www.firstfarminn.com
E-mail: info@firstfarminn.com
Directions: I-25 Exit 11, right one mile, right one mile. **Facilities:** 6 stalls, 3-12x15, 1-30x24 indoor. Feed/hay and trailer parking available. 6 acres or 1/2-acre board and Australian fencing Horses must have coggins and all vaccinations. **Rates:** $15-25 per night. **Accommodations:** Bed and Breakfast on site.

CLEVELAND
Pinehaven Stables Phone: 440-235-3200
Malcolm Cole
7611 Lewis Road [44138] **Directions:** 2 miles from I-480 & 4 miles from I-71. Call for directions. **Facilities:** 2 indoor stalls, 1/2 acre turnout, 55' x 120' turnout, feed/hay & trailer parking. Call for reservation. **Rates:** $10 per night; $70 per week. **Accommodations:** Red Roof Inn & Holiday Inn nearby.

COLUMBUS
Liberty Farm Phone: 614-279-0346
Kathryn Osborn
2620 Fisher Road [43204] **Directions:** 1 mile off of I-70 on west side of Columbus. Call for directions. **Facilities:** 30 indoor 12' x 12' box stalls, grooming stall, indoor arena, outdoor ring, 8 acres of pasture, full jump course on total of 17 acres. Training of horses & riders in hunt seat. Buying & selling of hunter/jumpers. Call for reservation. **Rates:** $20 per night; $77 per week. **Accommodations:** Motels 5 miles from stable.

DUBLIN
Dublin Stables Phone: 614-764-4643
Ginette Feasel
3910 Summit View Rd. W. [43016] **Directions:** 1.5 miles north of 270, off exit 20, Sawmill exit. Turn left at Summit View Rd.W. light, 4th drive on right. **Facilities:** Fortynine 8x10 to 10x16 indoor stalls. 60'x120' indoor arena, lighted 110'x250' outdoor arena, six pastures, 4+ acres and 2+ acres. Timothy/grass, 10% sweet feed, trailer parking available. **Rates:** $15 per day $75 per week. **Accommodations:** Please call.

FINDLAY
Lazy Creek Horse Farm Phone: 419-427-2400
Chris Lyon
11355 Hancock County Road [45840] **Directions:** Exit 23 off of I-75. Call for directions. **Facilities:** 17 indoor stalls, 4 holding pens, 3 pasture areas, indoor arena, feed/hay & trailer parking available. Training of horses & riders in English & Western. **Rates:** $10 per night. **Accommodations:** Motels 3 miles from stable.

OHIO PAGE 175

ALL OF OUR STABLES REQUIRE CURRENT NEG. COGGINS, CURRENT HEALTH PAPERS, & OWNERSHIP PAPERS.

FREDERICKTOWN
✱ Heartland Country Resort & Stables   Phone: 419-768-9100
Dorene Henschen (owner), Ben Daniel (mgr.)   or: 419-768-9300
2994 Township Road 190 [43019] **Directions:** Exit 151 off I-71, east on 95 for 2 miles; in Chesterville, south on 314 for 2 miles; left (east) on County Road 179 to intersection of Township Road 190. **Facilities:** 10 wood indoor 12' x 12' stalls some with outdoor turnouts; 20 acres of pasture, indoor and outdoor arenas, feed/hay, trailer parking. Trail riding, lessons in Western, English, hunt seat, barrel racing, or calf roping; horses available to rent with certified guides and instructors. Numerous activities available at award-winning resort. Call for reservations. **Rates:** $15 per night. **Accommodations:** B&B on premises.

HILLIARD
Country Squire Farms   Phone: 614-529-0055
Kathy Morgan
3687 Alton & Darby Creek Road [43026] **Directions:** 3 miles from I-70. Call for directions. **Facilities:** 9 indoor stalls, 3-4 acre pasture, seven 1/2- to 1-acre pastures, indoor arena, indoor exercise track, 2 outdoor arenas with jumps, 1/2-mile race track, plus 140 acres for riding. Call for reservation. **Rates:** $15 per night; $75 per week. **Accommodations:** Red Roof Inn in Hilliard, 3 miles from stable; Comfort Inn in Columbus, 3 miles from stable.

JACKSON
Henderson's Arena Complex   Phone 740-988-4700
Jerry L. Henderson
800 Van Fossan Road, County Road 23 [45640] **Directions:** On County Rd. 23, service-type road parallel to east/west 4-lane Hwy SR 32, almost midway to north/south 4-lane Hwy SRs 35 & 23. **Facilities:** 60 indoor 10' x 12' stalls aluminum fronts concrete aisle, 2 faucets in barns, two 130' x 250' riding rings, feed/hay available, trailer parking, camper hook-ups. Western wear store on grounds; saddle shop less than 3 miles away. **Rates:** $10 per night. **Accommodations:** Days Inn, Comfort Inn in Jackson, 10 miles from stable.

JACKSONTOWN
Beechwood Forest   Phone: 800-317-5157
Jeff Edwards
10050 Jacksontown Road [43030] **Directions:** 300 feet north of I-70, on SR 13. **Facilities:** 8 indoor box stalls, pasture/turnout area, trailer parking. Stable along with a 60-site RV park with full hook-ups. **Rates:** $15 per night. **Accommodations:** In addition to RV park, 2 motels within 4 miles.

OHIO

ALL OF OUR STABLES REQUIRE CURRENT NEG. COGGINS, CURRENT HEALTH PAPERS, & OWNERSHIP PAPERS.

LUCASVILLE
Silverstone Farm Phone: 937-259-5919
Tom Bombolis
735 Cook Road [45648] **Directions:** Located off of Rt. 23 N. Call for directions. **Facilities:** 25 indoor box stalls, 3 turnout paddocks, indoor arena, outdoor ring, wash racks, 2 miles of trails. Trains & shows horses & riders. Call for availability & reservations. **Rates:** $15 per night. **Accommodations:** Motels within 5 miles of stable.

MOUNT VERNON
Giddy Up Stables Phone: 740-397-6237
Keitha Smith   Web: www.members.tripod.com/piderwoman
8980 Tucker Road [43050] **Directions:** Visit website for directions. **Facilities:** Eleven 12x12 or 12x14 indoor stalls, feed/hay available, trailer parking, 5 pastures of various sizes. roundpen. Small quiet facility located on 16 acres in the pretty rolling hills southwest of Mt. Vernon. We cater to the individual needs of each horse. **Rates:** $15 per night. **Accommodations:** Mt. Vernon Inn and Curtis Hotel 6 miles from stable.

NEW LEBANON
Nickel & Dime Stable Phone: 513-687-3632
Carolyn Palmer
10582 Mile Road [45345] **Directions:** From I-70W take Rt. 49 Exit; south on Hoke Rd.; go to dead end, right on Westbrooke to STOP sign; left on Diamond Mill Rd.; cross Rt. 35 & take right on next street which is Mile Rd. **Facilities:** 6 indoor stalls, three 1/2-3 acres pasture/turnout, 1/2 acre pasture with separate barn, riding trails nearby. Dealer for UltraGuard vinyl fencing. All notices of arrivals made by 5 P.M. **Rates:** $15 per night. **Accommodations:** Camper possibly available at $15 per night; camping on premises for $5 per night; Days Inn in Brookville, 10 miles from stable.

SEVILLE
✶ **Rocking H Ranch** Phone/Fax: 330-334-5466
Christopher & Lori Minnich
3737 Seville Road [44273] **Directions:** At junction of I-71 and I-76; Exit 2, 15 miles south of I-271. Call for directions. **Facilities:** 12 - 12' x 12' stalls, 2 holding pens, 6 pastures, feed/hay and trailer parking available. Fishing pond; dogs welcome. Reservations required. **Rates:** $15 per night. **Accommodations:** Fully furnished room with private bath available.

ALL OF OUR STABLES REQUIRE CURRENT NEG. COGGINS, CURRENT HEALTH PAPERS, & OWNERSHIP PAPERS.

STREETSBORO
<u>Sahbra Farms</u>   Phone: 440-626-2040
David Gross
8261 Diagonal Road [44241] Directions: Exit 13 off of I-80, 1.5 miles to farm. Call for directions. **Facilities:** 215 indoor stalls, 30 outside paddocks, 1/2 mile race track. This is a beautiful breeding & training center for standardbreds on 400 acres. Call for availability & reservation & rates for use of training center. **Rates:** $14 per night for stall and turnout. **Accommodations:** 1 mile away.

SWANTON
<u>Post and Rail Stables</u>   Phone: 419-826-9934
Bonnie Cicora
10362 State Route 64 [43558] Directions: Call for directions. **Facilities:** 43 indoor stalls, many large pasture areas and round pens, feed/hay & trailer parking available. Arrival must be before midnight and reservations required. This is a complete training facility. **Rates:** $25 per night; $100 per week. **Accommodations:** Cross Country in Toledo and Fairfield Inn in Toledo, 10 minutes away.

WEST ALEXANDRIA
✶ <u>Surecare Farm</u>   Phone: 937-839-4186
Scott & Barb Stockslager   E-mail: surecare@infinet.com
7089 State Route 35 East [45381] Directions: Call for Directions. **Facilities:** 4- 10' x 20' indoor box stalls. Feed/Hay, trailer parking available. Bed and continental breakfast for guests *and* horses! Easy access for truck and trailer. Located 8 miles from I-70 on State Route 35. **Rates:** $90/night for horses, owner, and dog. **Accommodations:** Small apartment with full bath and queen size bed.

PAGE 178 **OKLAHOMA**

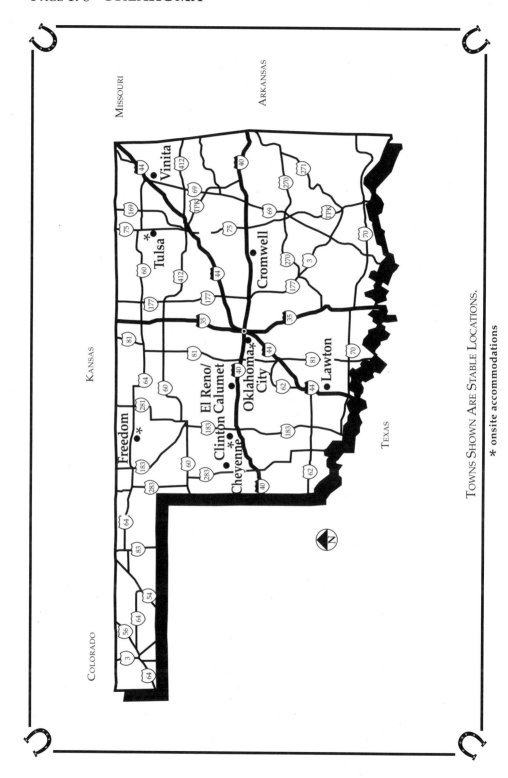

TOWNS SHOWN ARE STABLE LOCATIONS.

* onsite accommodations

OKLAHOMA Page 179

ALL OF OUR STABLES REQUIRE CURRENT NEG. COGGINS, CURRENT HEALTH PAPERS, & OWNERSHIP PAPERS.

CHEYENNE
✱ Coyote Hills Guest Ranch   Phone: 580-497-3931
Kass Nickels   Web: www.coyotehillsguestranch.com
E-mail: coyotehr@aol.com
P.O. Box 99 [73628] **Directions:** Exit 20 off I-40 at Sayre, Hwy 283 north 28 miles to Cheyenne; take Hwy 47 west 4 miles, then 2 miles north, then 2 miles west. **Facilities:** 8 indoor 12' x 12' stalls, 24' x 12' outside barns with 32' x 32' runs, 150' x 60' paddocks, full size outdoor arena and round pen, feed/hay, trailer parking with RV hookups. Horseback riding. Located in the beautiful red hills of Roger Mills County. please Call for reservations. **Rates:** $20 per night. **Accommodations:** 4-star Territorial Hotel on ranch with meals served; private rooms with bath, heated and air conditioned.

CLINTON
JR'S Western Horse Motel   Phone: 580-323-5888
M. L. (JR.) Richardson   Phone: 580-323-1588
1710 Neptune Drive [73601-9722] **Directions:** Turn south on Neptune Drive at 65A exit. 3/4 mile to 1710 Neptune Drive. **Facilities:** Seven pens with sheds approximately 40' X 60', <u>NO Box Stalls.</u> No feed or hay available. Trailer parking available. Back yard facilities, friendly "at home" atmosphere. **Rates:** $15/$20 per night. **Accommodations:** Four nice motels within a mile of stable and R.V. park 1/4 mile.

CROMWELL
Four Winds Ranch   Phone: 405-944-1180
Michelle Richardson   E-mail: fourwindsranch@yahoo.com
PO Box 85 (74837)   Web: www.fourwindsranch.net
Directions: Exit 212 off I-40. Food, fuel, diesel mechanic available at exit. Call for directions. **Facilities:** All new facility consists of 10 indoor stalls all rubber matted, vary in size, 2 large pastures. Feed/hay available, trailer parking available. Public trail riding only 20 mins away. Staff provides full care. **Rates:** $12-20 depending on size of stall. $100 weekly rate. **Accommodations:** 3 RV spaces at ranch, Motor Lodge 12 miles away in Okemah, Seminole Inn 15 miles away in Seminole.

EL RENO
Little Bit Farm   Phone: 405-262-7504
David & T.J. Meschberger
Route 2, Box 43, Calumet [73014] **Directions:** From I-40: Take Exit 119; go about 1/2 mile north to Ft. Reno crossover; take left & go one mile west; take right & go 1/2 mile north; white fence on left is stable. **Facilities:** 20 indoor stalls, 150' x 150' turnout, no feed/hay, trailer parking on premises and can accommodate big rigs. Electric hook-up for camper. Vet nearby. Reservations preferred. **Rates:** $15 per night. **Accommodations:** Best Western Inn in El Reno, 5 miles from stable.

OKLAHOMA

ALL OF OUR STABLES REQUIRE CURRENT NEG. COGGINS, CURRENT HEALTH PAPERS, & OWNERSHIP PAPERS.

FREEDOM
✶ **Cedar Canyon Lodge** Phone: 580-621-3257/3258
Anita Rennebohm & Marvin Nixon   Web: www.cedarcanyonlodge.net
Rt 1 Box 31 B [73842] **Directions:** Hwy 50, NW corner of OK, Woodward County located between Alva & Woodland, 19 miles N of Moorland, OK **Facilities:** 6 outdoor arenas with pens, feed/hay available, trailer parking, 40 acres of pasture, trail riding on 3500 acre ranch. Please visit our website www.cedarcanyonlodge.net for information on lodging, restaurant and RV parking. **Rates:** $15 per night, $125 per week. **Accommodations:** Cedar Canyon Lodge at facility.

LAWTON
Red Horse Ranch, Arena & Stables Phone: 405-492-4856
Betty & Evonne Beaty
7323 NW Wolfe Road [73505] **Directions:** I-44 to Hwy 49; go west 4 miles to Hwy 58 & go north 6 miles; big yellow metal building on east side of Hwy 58. (Call for map.) **Facilities:** 24 indoor stalls at least 12' x 12', one 10-acre, three 2-acre, & three 1-acre pasture/turnout areas; 100' x 225' indoor arena; outdoor arena; indoor heated wash rack, hot walker, 80' x 45' round pens; large riding area available; less than 2 miles from mountains & lake. Oats, omolene, & prairie hay available. Trailer parking. Request that horse owners provide own feed buckets. Call for reservation. **Rates:** $12 per night; $60 per week. **Accommodations:** Howard Johnson's 17 miles away in Lawton (1-800-446-4656).

OKLAHOMA CITY
AAA Horse Motel & Thoroughbred Training Phone: 405-771-3472
R.L. Odom
6920 N.E. 63rd [73141] **Directions:** Call for directions 30 minutes to 1 hour in advance. **Facilities:** 20 indoor stalls, five separate 2-6 acre paddocks, 1/2-mile racetrack, feed/hay & trailer parking available. Open 24 hours. Training, racing, & sales of thoroughbred horses. **Rates:** $15 per night; $10 for 2nd and more nights. **Accommodations:** Holiday Inn & Best Western 3 miles away.

Ed Cook Phone: 405-634-1787
8400 S. Walker [73139] **Directions:** Call for directions. **Facilities:** 7 indoor 14' x 14' stalls, four holding pens, 5-acre pasture, feed/hay & trailer parking available. **Rates:** $15 per night; $75 per week. **Accommodations:** Many motels 1/2 mile east of stable.

✶ **Robinson's Barn-Bed-Breakfast** Phone: 877-733-2443
Kenn & Donna Robinson   Web: http://home.earthlink.net/~horseldy
11200 SE 44 Place [73150] **Directions:** I-40 & Anderson Road. **Facilities:** 8 indoor concrete 10' x 12' stalls with pipe runs, 100' x 100' turnout, large riding arena, feed/hay, trailer parking; kennel and RV hook-ups ($20 per night) available. **Rates:** $15 per night. Medicinal cob webs, no charge. **Accommodations:** Two bed & continental breakfast in barn, rooms with full bath, heat/air in each room. One room with twin beds, one room with queen bed, $45 per night. 20 minutes from State Fair grounds.

OKLAHOMA PAGE 181

ALL OF OUR STABLES REQUIRE CURRENT NEG. COGGINS, CURRENT HEALTH PAPERS, & OWNERSHIP PAPERS.

TULSA
* Bar "B" Farms — Phone: 918-645-2876
Suzanne Owen — Phone: 918-645-2876
Ken Beets — Phone: 918-232-7077
E-mail: banoahnjames@aol.com
13201 S. Lewis St [74101] **Directions:** I-44 to Hwy 75 go south to 141st St. Turn east to Lewis St. Turn north go 3/4 of a mile. We are on the east side of Rd. Look for sign, silver barns back off road. **Facilities:** 11 14x14 indoor stalls, 8 outdoor paddocks. Prairie hay and alfalfa. Trailer parking available. 1-3 acre paddock. Boarding facility, jumping, English and Western Riding lessons. Close to Turkey Mountain many miles of trials. Neg, coggins, West Nile vaccination and health papers required. **Rates:** $10 per horse per night, $55 per horse per week. **Accommodations:** Electric hook-up $10 per night. Queen size bed available in bunkhouse on property $10 per person. Best Western 5 miles (918-322-3180)

Bar "B" Farms

- 11- 4x14 Indoor Stalls
- 8 Outdoor Paddocks
- Boarding & Jumping Facility
- English & Western Riding Lessons
- Trailer Parking Available
- Miles of Trials Near Turkey Mt.

918-645-2876

13201 S. Lewis St, Tulsa, OK (74101)

OKLAHOMA

ALL OF OUR STABLES REQUIRE CURRENT NEG. COGGINS, CURRENT HEALTH PAPERS, & OWNERSHIP PAPERS.

VINITA
Double D Ranch Phone: 918-256-6268
Paul & Deanne Brown E-mail: Paulndee@NEOK.com
447954 E Hwy 60 (74301) Directions: 7 miles off I-44. Take Afton exit 302 on Hwy 60/69 for 10 miles or take Vinita exit 289 on Hwy 60/69 for 7 miles. **Facilities:** 80-acre ranch. 40 indoor stalls, turnout pens, fee/hay, indoor riding arena indoor wash rack and trailer parking on premises. Camper parking OK if self-contained. Ranch breeds and sells pure Polish Arabians. **Rates:** $15 per horse, extended stays available. **Accommodations:** Motels within 7 miles.

NOTES AND REMINDERS

Page 184 OREGON

Towns Shown Are Stable Locations.

* onsite accommodations

OREGON Page 185

ALL OF OUR STABLES REQUIRE CURRENT NEG, COGGINS, CURRENT HEALTH PAPERS, & OWNERSHIP PAPERS.

ALBANY
Springhill Boarding Stables Phone: 541-928-8943
Liz & Jerry Couzin
5368 Springhill Drive [97321] **Directions:** Located in North Albany. Call for directions. **Facilities:** 33 indoor stalls, 50' x 150' and 60' x 70' indoor arenas, 80' x 150' outdoor arena, 66' round pen, hot walker, feed/hay & trailer parking on premises. Horses for sale & lease. Training of horses and riders available in all disciplines at all levels. Standing-at-stud: Champion Khemosabi ++++ Son, "Khareen" and Champion Bey Shah Grandson, "SS Shahzahm." Both stallions are throwing beautiful, correct, winning offspring with great minds and fabulous dispositions. Call for reservations or more info. **Rates:** $15 per night with discount for multiple horses. **Accommodations:** Best Western, Holiday Inn, Super 8, Comfort Inn located 5-7 miles from stable.

ASHLAND
B-G Valley View Stables Phone: 541-482-1772
Virginia J. Blair
263 Wilson Road [97520] **Directions:** Exit 19 off of I-5: from the south, exit N. Valley View Rd.; turn right & go 1 mile then turn right onto Wilson Road. Stable is on left, 1,000 ft. from road. S.O.S. sign on driveway. **Facilities:** 8 indoor 12' x 12' stalls, 70' x 120' indoor ring, seven 12' x 32' runs, big turnaround area for trailers, hot & cold horse shower, grain/hay furnished. No smoking. fire sensors. No loose dogs. Must be in by 10 P.M. Please make reservation. Will not return long distance calls. Must have current strangles, rhino, potomac fever, rabies, elw. sleeping sickness, tetanus shots. 42 years experience with horses. Trailer & camper parking available. **Rates:** $15 per night per horse. Full care. **Accommodations:** Best Western Heritage Inn (will take pets) 1 mile away.

COOS BAY
Family Four Stables Phone: 541-267-5301
James & Peggie Henriksen Fax: 541-269-7059
61676 Family Four Drive [97420] **Directions:** 7 miles south of Coos Bay on Hwy 101. Call for directions. **Facilities:** 50 indoor 12' x 12' double-walled stalls, feed/hay, trailer parking, 3-10-15 acre pastures, 25-15x35 outdoor paddocks, 100'x200' covered arena, 75x150 outdoor arena, 50x50 indoor lounging arena, hot walkers, round pen and washing facilities. Summer - guided trail rides and youth day camps, riding lessons and certified trainers available. Horses for rent or lease. Electric hookups for self-contained RV's. **Rates:** $15 per night, $10 per night weekly rate. **Accommodations:** Motels in Coos Bay, 7 miles away.

PAGE 186 OREGON

ALL OF OUR STABLES REQUIRE CURRENT NEG, COGGINS, CURRENT HEALTH PAPERS, & OWNERSHIP PAPERS.

DIAMOND LAKE
Diamond Lake Corrals   Phone: 541-793-3337
Wayne Watson
750 Diamond Lake Loop [97731] **Directions:** Call for directions; off Rte. 138 to Roseburg. **Facilities:** Five-stall barn, 14 corrals of various sizes, feed/hay, trailer parking available. Horse rentals, pack trips, hunting parties; 20 miles from Crater Lake. Open June-September. Space to pitch a tent or park a camper. **Rates:** $7.50 per night. **Accommodations:** Diamond Lake Resort 1/4 mile away.

HALSEY
Pioneer Stables   Phone: 541-369-3622
Willow Coberly   Fax: 541-369-3634
Web: pioneerstables.com   E-mail:overnight@pioneerstables.com
32744 Hwy 228 [97348] **Directions:** From I-5 take Exit 216, go west 1/2 mile to Falk Rd. Barn is on corner of Hwy 228 and Falk Rd. **Facilities:** 38 12x12 rubber matted indoor stalls, 60x156 indoor arena, 36x36 covered turnouts, several outdoor turnouts. Feed/hay and trailer parking available. **Rates:** $20-25 per horse per night. $125 weekly rate. **Accommodations:** Best Western at exit 216, 1/2 mile from stable.

LAGRANDE
Mavericks, Inc.   Phone: 541-963-3991
Maverick Inc. Riding Club
3608 2nd Street N. [97850] **Directions:** 100 yards off of I-84. Call for directions. **Facilities:** Ten 12' x 12' outdoor pens, large outdoor arena, no feed/hay available, trailer parking on premises. Open horse shows, team-penning, & team-roping events held at facility. Riding trails available. Call ahead. **Rates:** $5 per night. **Accommodations:** Greenwell Motel within 1 mile of stable.

LAPINE
J & B Acres   Phone: 541-536-5833
Barbara Roof   E-mail: jandbaines@coinet.com
52097 Pine Forest Drive [97739] **Directions:** Hwy 97 North of LaPine, take Burgess Rd. 2 1/2 miles to Pine Forest, go 3/4 mile. Call in advance for further information. **Facilities:** 3 outdoor 30' round pens, alfalfa/grass hay available, trailer parking. Nothing fancy, just outdoor paneled pens in a quiet neighborhood. **Rates:** 1st horse-$20 per night, 2nd horse-$15, 3rd horse-$10. **Accommodations:** Hidden Pines RV park 1/2 block away, Best Western in LaPine 3 miles away.

OREGON PAGE 187

ALL OF OUR STABLES REQUIRE CURRENT NEG, COGGINS, CURRENT HEALTH PAPERS, & OWNERSHIP PAPERS.

NEHALEM
<u>Pearl Creek Stables</u>　　　　　　　　　　　　Phone: 503-368-5267
Janice Woelfle
17150 Camp Four Road [97130] **Directions:** Call for directions. Hwy 53 off US 101, about 3 miles to second Camp Four Road sign. **Facilities:** 7 indoor 10' x 12' stalls, rubber matted, bedding provided, small corral for turnout, 50' x 60' indoor arena, trailer parking available. Nehalem Bay State Park 5 minutes away, trailer to Manzanita Beach. Dog obedience training. Call for reservations. **Rates:** $15 per night. **Accommodations:** Lodging within 5 miles; call for information. Camping available at state park.

OAKLAND
<u>Dodge Creek Stables</u>　　　　　　　　　　　Phone: 541-459-2609
Mary Bayard　　　　　　　　　E-mail: mary@americanblacksmith.com
Web: www.americanblacksmith.com/dodgecreek
3739 Hwy 138W [97462] **Directions:** Exit 136 (Sutherlin) off I-5, west on Hwy 138W for 3.5 miles, stable on right. **Facilities:** 21 - 12' x 12' and 12' x 14' box stalls with 24' runways, pasture/turnout in various sizes, huge indoor arena, wash rack, trails, jumps, outside lighting, feed/hay, trailer parking. Photos on website. Training and lessons in dressage, eventing, jumping. Emergency pick-up available. Caretaker on grounds. Electricity, water, shower available for camping. Call for reservations. **Rates:** $25 per night; $125 per week. Fee for late arrival. **Accommodations:** Campers welcome on premises. reasonable, nice motels at exit #136, 3.5 miles from stable.

ONTARIO
<u>Malheur County Fairground</u>　　　　　　　Phone: 541-889-3431 (days)
Janeen Kressly　　　　　　　　　　　　Evening: 541-823-2581 (John)
E-mail: mcfair@fmtc.com　　　　　　　　　　　　Cell: 541-823-2997
795 NW 9th Street [97914] **Directions:** Located 2 miles off of I-84. Exit 376A Call for directions. **Facilities:** 100 outside covered stalls, outdoor riding arena, rodeo-size paddock area, feed store nearby, trailer parking on premises. Open 24 hours with caretaker on site. Fairground hosts rodeos, gun shows, circus, and horse shows, auctions, annual car show and more. **Rates:** $10 per night, includes shavings. **Accommodations:** Holiday Inn & Best Weatern -2 miles away.

PENDLETON
✱ <u>NEIGH-bors (it helps if you say it with a whinny)</u>　Phone: 541-276-6737
Mary Alice Ridgway　　　　　　　　　　　E-mail: emayridg@UCI.net
543 NW 21st St [97801]　　　　　　　　　　Web: www.neigh-bors.com
Directions: Call for directions or see our website. **Facilities:** 7 stalls. 4 12'x12' indoor, 3 12'x12' outdoors. Grass/alfalfa available, short-term trailer parking available. 2 fields, 2 acres each. Health certificates and coggins required. **Rates:** $20 per night. $65 per night for condo. **Accommodations:** Stay with your horse in our A/C and modern condo at the barn, Bed & Barn, Bed & Breakfast. Oxford Suites, Red Lion, Holiday Inn Express, Motel 6, many more.

ALL OF OUR STABLES REQUIRE CURRENT NEG. COGGINS, CURRENT HEALTH PAPERS, & OWNERSHIP PAPERS.

SANDY
Burnt Spur Ranch Phone: 503-668-9716
Linda Keeter & John Keeter Web: www.1010design.com/burntspur
E-mail: lindank1@ipns.com
42100 SE Locksmith Lane [97055] **Directions:** 2 miles east of Sandy, South at Shorty's Corner onto Firwood Road, go 2 miles and turn left onto Locksmith Lane. **Facilities:** 7 indoor stalls, 6 - 12' x 24' covered outdoor pens, hot walker, 110' x 120' outdoor arena, feed/hay, trailer parking. Professional horse trainer on premises. Overnight camper hook-ups. Less than 1 hour from skiing, fishing, trails, etc. Call for reservations. **Rates:** $15 per night. **Accommodations:** B&B within 3 miles. Motel 2 miles. Several Restaurants and Lounges 2 miles.

SISTERS
Black Butte Stables Phone: 1-800-743-3035
Sandra Herman, Owner
Mike Elmore, Manager
P.O. Box 418 [97759] **Directions:** 7.5 miles west of Sisters on Hwy 20 on Black Butte Ranch, behind the general store. **Facilities:** 10 indoor 12' x 12' stalls, 8 outdoor 10' x 30' paddocks, feed/hay available, trailer parking. Primarily a guided horseback riding and wilderness pack station. Located in foothills of Three Sisters Wilderness Area. Exceptional riding opportunities - waterfalls, creeks, alpine meadows, natural crater-formed lake. Call for reservations 2-3 days in advance in off season; 2 weeks in advance during summer. **Rates:** $7.50 per night self-paddock; $50 per week self-paddock. **Accommodations:** Best Western, Comfort Inn in Sisters, 7.5 miles away.

SPRAGUE RIVER
✻ **Cracker Jack Ranch** Phone: 541-533-3400
Jeanee & Lew Conner Fax: 541-533-3418
E-mail: crackrjk@cvc.net Web: connerquarterhorses.com
33106 Klamath Forest Dr [97639] **Directions:** Please call. **Facilities:** 8 12'x12' stalls. Indoor sand arena. Feed/hay, trailer parking available. Pipe corrals 32'x32'. 100 acres next to Winema National Forest with miles of moutain trails to ride. Also dirt roads for those who may want to drive their horses. Four Quarter Horse Stallions for breeding. **Rates:** Varies with accommodations, please call. **Accommodations:** We are a Bed & Barn which consists of 1 bedroom cottage w/complete facilities sleeps 5-6. One bedroom trailer sleeps 4. Many motels in Klamath Falls.

OREGON

ALL OF OUR STABLES REQUIRE CURRENT NEG. COGGINS, CURRENT HEALTH PAPERS, & OWNERSHIP PAPERS.

SUNNY VALLEY

T & R Stables/Grant's Pass/Sunny Valley KOA Phone: 541-474-2516
Tom & Rose Johnston   E-mail: tandrstables@aol.com
359 Old Stage Road [97497] **Directions:** From North take exit 71; follow signs to KOA Campground or, from the South, take Exit 71; go under overpass, take the first left and drive past and behind to 359 Old Stage Rd. to a 2 story red house. **Facilities:** 7-12'x 14' indoor stalls with paddocks on 3 acres. No feed or hay available. Horse trailer parking available. No stallions. **Rates:** $14 per night, weekly rate available upon request. **Accommodations:** Motels 12 miles from stable.

THE DALLES

Fort Dalles Days Rodeo Association   Phone: 541-296-6817
Steve Hunt, President   Night: 541-298-8671
Bargeway Road [97058] **Directions:** 1 mile from I-84. Call for directions. **Facilities:** 12 stock pens on 14 acres, 200' x 300' outdoor arena with 3,000 seats & concession stands. Pro & junior rodeos held at facility. Year-round caretaker on site. Hughes Feed & Grain & tack store nearby. K&H Specialties in The Dalles is a full-service repair shop that can fix trucks & trailers with mechanical problems. **Rates:** No charge but donations gladly accepted. **Accommodations:** Quality Inn & Cousins Restaurant 1 mile from rodeo grounds.

PAGE 190 **PENNSYLVANIA**

TOWNS SHOWN ARE STABLE LOCATIONS.

* onsite accommodations

PENNSYLVANIA PAGE 191

ALL OF OUR STABLES REQUIRE CURRENT NEG. COGGINS, CURRENT HEALTH PAPERS, & OWNERSHIP PAPERS.

ALTOONA
Sinking Valley Stables
Chris Beaver
R. D. 3, Box 255 [16601]

Phone: 814-944-3241
Barn: 814-944-7063

Directions: Rt. 220 N to Altoona from Pittsburgh; 17th St. Exit right at light; go through 1 light & at 2nd light turn right onto Kettle St.; go 7 miles & farm is on left. **Facilities:** 5-18 indoor 10' x 12' stalls; 50' x 50' paddocks; timothy, alfalfa, & orchard grass available. Trailer parking. **Rates:** $10 per night - $8 per night if more than 1 horse; $20 for weekends. **Accommodations:** Holiday Inn, Days Inn, & EconoLodge all in Altoona, 10 minutes from farm.

BETHEL
* Windy Ridge Farm
Judy Reggio
E-mail: travel@windyridgefarm.com

Phone: 717-933-5888
Web: www.windyridgefarm.com

401 Swope Road [19507] Directions: 2 miles north of I-78, between Exits 10 & 13. 30 minutes from Hershey Park. Located in Amish Country, a vacation destination. Call for directions. **Facilities:** 5 large, airy indoor stalls, three 1/4-acre parcels of grass, one 3-acre board-fenced pasture area with run-in, feed/hay & trailer parking. Farm is a breeding facility for Dutch warmbloods by a national top stallion. Nationally recognized young warmbloods for sale. Devon winners 2 years running. Must have rhino shots. **Rates:** $25 per horse, per night. Reduced when staying at our B&B. **Accommodations:** Guest rooms available in new home on a beautiful farm nestled at the foot of the Blue Mountains.

BLOOMSBURG
Engelwood Paint Horses &
Animal Chiropractic Clinic
Neil & Janine Engelman

House Phone: 570-387-0510
Barn Phone: 570-387-6655

35 Horse Farm Road [17815] Directions: From I-80: take the Lightstreet exit 236, From I-81: Take I-81 to I-80 take exit 236. Turn right at ramp stop sign, take 1st right onto Sawmill rd, drive 1.5 miles. **Facilities:** 10+ 12'x12' horse safe stalls w/sliding doors, many types of feed/hay available, trailer parking available, individual 1-acre turnouts. private paint Horse Farm w/ four rail wooden fence on 100+ acres. Indoor and outdoor areas. Animal Chiropractic services available, Please Bring own buckets. **Rates:** $25 per night, weekly rates available apon request. **Accommodations:** Several motels and restaurants within 2-5 miles of the facility.

BROOKVILLE
Seneca Trial Horses
John McAninch & Hollie Nelson

Phone: 814-849-8135

RR 8 P.O. Box 64 [15825] Directions: I-80 exit 13, 2 miles south on Rt 28, turn left on Seneca Trial Rd, farm is 1 mile on left. **Facilities:** 10 stalls, 10' x 12' open air / rubber mats. Several pastures available, 1-3 acres. 60' x 60' indoor arena, several miles of scenic wooded trails, mare and foal stall available. **Rates:** $15 per night. **Accommodations:** Brookville at exit 13 I-80, 3 miles from stable, Super 8, Days Inn, Holiday Inn. Also B & B, G. Barrier House.

PAGE 192 PENNSYLVANIA

ALL OF OUR STABLES REQUIRE CURRENT NEG. COGGINS, CURRENT HEALTH PAPERS, & OWNERSHIP PAPERS.

BUCKINGHAM
Mill Creek Farm Enterprises, Inc.   Phone: 215-794-3121
James R. Brame, Sr.
2348 Quarry Road [18912] **Directions:** I-95 Exit 30: Take Rt. 332 west for 3 miles; take Rt. 413 north 8 miles to property. **Facilities:** 13 large 14' x 14' indoor box stalls, 10 acres of pasture/turnout, feed/hay & trailer parking available. Coggins required. Also swimming pool with spa/jacuzzi, tennis court, etc. **Rates:** $25 per night **Accommodations:** Bed & Breakfast on premises. Rooms are $100 per night, midweek, including breakfast.

CARLISLE
✶ **Pheasant Field Bed & Breakfast**   Phone: 717-258-0717
Denise (Dee) Fegan   Web: pheasantfield.com
E-mail: stay@pheasantfield.com
150 Hickorytown Road [17013] **Directions:** From I-76 and I-81: Southeast on S. Middlesex Road, left on Ridge Drive, right on Hickorytown Road. Within 1 hour of Hershey, Harrisburg, Gettysburg. **Facilities:** 12 indoor 15' x 15' stalls, 4 outdoor run-in sheds, 4 acres of pasture, feed/hay available, trailer parking. **Rates:** $20 per night. **Accommodations:** Bed & Breakfast on site. Non-smoking.

COVINGTON
Tanglewood Camping   Phone: 570-549-8299
Mark Paone   Web: tanglewoodcamping.com
E-mail: tnglewod@ptd.net
RR1 Box 64A [16917] **Directions:** Call for directions. **Facilities:** water and electric campsites available $28 per night, with plenty of room to camp with horse at your site. Bring portable corrals or picket lines. **Rates:** $3.00 per horse. **Accommodation:** Campground

DANVILLE
✶ **Circle G Riding Stable, Inc.**   Phone: 570-275-3099
Charles R. Gordner
2903 Bloom Road [17821] **Directions:** Call for directions. **Facilities:** 6 indoor stalls, feed/hay & trailer parking available. No smoking in barns. Call for reservation. **Rates:** $15 per night; $75 per week. **Accommodations:** Bed and Breakfast on site $85-$115 dbl includes full breakfast. Red Roof Inn & Days Inn in Danville, 20 minutes from stable.

DUSHORE
Drake Hollow Stables   Phone: 570-928-7101
Robert Brown
RR4, Box 4307 [18614] **Directions:** Call for directions. Off of Exit 34 on I-80. **Facilities:** 12 indoor 12' x 12' stalls, small pasture, 60' round pen, 100' x 150' arena, 1-acre turnout, feed/hay, trailer parking. Call for reservation. **Rates:** $20 per night. **Accommodations:** Motels in Dushore, 5 miles from stable.

PENNSYLVANIA Page 193

ALL OF OUR STABLES REQUIRE CURRENT NEG. COGGINS, CURRENT HEALTH PAPERS, & OWNERSHIP PAPERS.

EIGHTY FOUR
Gil-Mar Stables Phone: 724-942-4354
Gary Stegenga, Manager
425 Ross Road [15330] **Directions:** Canonsburg Exit off of I-79. Stable is 6 miles from exit. **Facilities:** 20-stall barn, 120' x 120' paddock, outside pasture, outside ring, feed/hay & trailer parking available. Call for reservation. **Rates:** $15 per night. **Accommodations:** Motels 10 miles from stable.

GIBSONIA
✱ **Sun & Cricket Farm B & B** Phone: 724-444-6300
John & Tara Bradley-Steck
1 Tara Lane [15044-5507] **Directions:** 5 miles NE of I-76 (PA turnpike) at Exit 4; 12 miles east of I-79 at Warrendale Exit; 50 miles south of I-80. Call for further directions. **Facilities:** 5 indoor 10' x 12' box stalls, 1-indoor 12' x 12' box stall, 1/4-acre grassy pasture with split-rail fencing. Extensive trails. Other pets with prior approval only. Deposit required. No arrivals after 10 P.M. No smoking. Guests with horses must stay at B & B. Reservation required. **Rates:** $20 for stall; $15 for pasture per night. **Accommodations:** Carriage house with queen bed, primitive antiques, folk art, quilts, full bath, private porch, $120. Three-room log cabin with king bed or two twins, fireplace, atrium & library, $130. Discounts for two-night stays. Major credit cards accepted.

GRANTVILLE
Centaur Farm Phone: 717-865-5501
Edwin Stopherd or Ronald Gerstner
6500 Mountain Road (Rt 443) [17028] **Directions:** Exit 28 off of I-81: Take left off exit if eastbound - right if westbound; go to "T" & take right; go approx. 2.7 mi. & farm is on right. Green barn in rear. 18 wheelers OK. 1.6 mi. past Penn National Race Course. **Facilities:** 30 indoor stalls in totally enclosed barn, outdoor sheds, 50' x 50' paddocks, five 4-acre paddocks, 24-hr. vet available, lay-up & broodmare care, long- or short-term boarding available. Standing-at-stud: Thoroughbred "Dr. Koch" by "In Reality"; Saddlebred "Centaur's Fame and Fortune" by "Chief of Graystone." Any size trailer welcome. Call for availability. Payment in advance. **Rates:** $15 per night; $100 per week. Any special care is extra charge. **Accommodations:** Holiday Inn & Budget Motel in Grantville, 3 miles from stable.

HONESDALE
✱ **Triple W Riding Stable** Phone: 570-226-2620
Kevin Waller   Walter Stolle, mgr.
RR 2, Box 1540 [18431] **Directions:** 30 miles east of Scranton, 3 miles from Holly. Near Rt. 6. Call for directions. **Facilities:** 20 indoor box stalls, 4 fenced fields, indoor arena, wash rack, viewing room. Corrective shoeing & dentistry work available. Horse training, lessons, & sales, particularly Arabians & Appaloosas. Hay & sleigh rides. Call for reservation. **Rates:** $15-$20 per night. **Accommodations:** Bed & Breakfast on premises has 10 rooms with double occupancy. Also overnight camping.

ALL OF OUR STABLES REQUIRE CURRENT NEG. COGGINS, CURRENT HEALTH PAPERS, & OWNERSHIP PAPERS.

KYLERTOWN
Terri's Stable Phone: 814-345-6940
Terri Fisch
487 Second Street [16847] **Directions:** 1/2 mile north of Exit 21 off I-80. **Facilities:** 25 indoor stalls, hot & cold water, 3 wooden-fence turnout paddocks, indoor arena, feed/hay & trailer parking available. Training & breeding of quarter horses at stable. Call ahead for reservation. **Rates:** $15 per night. **Accommodations:** Stop 21 Motel 1/2 mile.

LANCASTER
Foxfield Farm Phone: 717-484-2250
Susan Walmer, Owner
230 Holtzman Road [17569] **Directions:** Exit 21 off of I-76. Call for directions. **Facilities:** 45 box stalls, indoor arena, paddocks, outdoor riding rings, feed/hay & trailer parking available. Call for reservation. **Rates:** $20 per night. **Accommodations:** Motels within 1 mile of stable.

NEWPORT
Windy Ridge Acres Phone: 717-567-7457
Bob Martin
3791 Middleridge Road [17074] **Directions:** Located off of Rt. 322. Call for directions. **Facilities:** 6 indoor box stalls, 4 tie stalls, 100' x 200' outdoor ring, 50' x 100' indoor ring, trail riding on 125 acres of wooded trails. English & Western lessons given at all levels. Also Clydesdale wagon rides available. Call for reservation. **Rates:** $15 per night. **Accommodations:** Motels within 5 miles of stable.

ZIONSVILLE
Horseman's Hollow Equestrian Center Phone: 215-541-4363
Elizabeth Lafrenz
8300 School House Lane [18092] **Directions:** Off PA. Turnpike, take Exit 32. Exit toll booth, right turn onto PA. Rt 663. Go 4/10 mile to Spinnerstown Rd. Turn right and go straight for 2 miles. At 2nd stop sign continue straight onto Orchard Rd. Go 6/10 mile to stop sign. Left onto School House Lane. 1 Mile to Horseman's Hollow on right. **Facilities:** 8 - 12' X 12' stalls. Tongue and groove boards. Hay & feed included. Trailer parking available. 12 acres of pasture. Boarding, lessons and 80' X 152' indoor arena. **Rates:** $15 per night. **Accommodations:** Econolodge and Rodeway Inn in Quakertown 4 miles away.

TRY A NEW & DIFFERENT TRAIL !
TAKE A TOUR OF PENNSYLVANIA'S HORSE TRAILS.

RIDE PENNSYLVANIA HORSE TRAILS©

Public horseback riding trails in Pennsylvania.
Includes great day rides, weekend get-aways, and week-long destinations.

Not just an ordinary guide book! Book includes color photos, area history, emergency vet & farrier contacts, nearby stables, trail location, length, level of difficulty, terrain, contact numbers, directions, description, equestrian camping accommodations and much more!

Order by credit card via website:

www.ridepennsylvania.com or
www.patrail.com

or mail check to :
Hit The Trail Publications, LLC at P.O. Box 970, Cherryville, PA 18035.
Don't forget to include delivery name, address, phone, email address.

Ride Pennsylvania Horse Trails, Part I
The Eastern Half of Pennsylvania

Cost of Book: **$22.95** x quantity of books $ _____

Sales Tax: PA residents add 6% sales tax ($1.38 x quantity)   + _____

Shipping: $4.00 first book, $2.00 each additional book   + _____

Total Enclosed :(For PA residents, 1 book, total would be $28.33) = $ _____

Ride Pennsylvania Horse Trails, Part II (NEW!)
The Western Half of Pennsylvania

Cost of Book: **$23.95** x quantity of books $ _____

Sales Tax: PA residents add 6% sales tax ($1.44 x quantity)   + _____

Shipping: $4.00 first book, $2.00 each additional book   + _____

Total Enclosed :(For PA residents, 1 book, total would be $29.39) = $ _____

PAGE 196 RHODE ISLAND

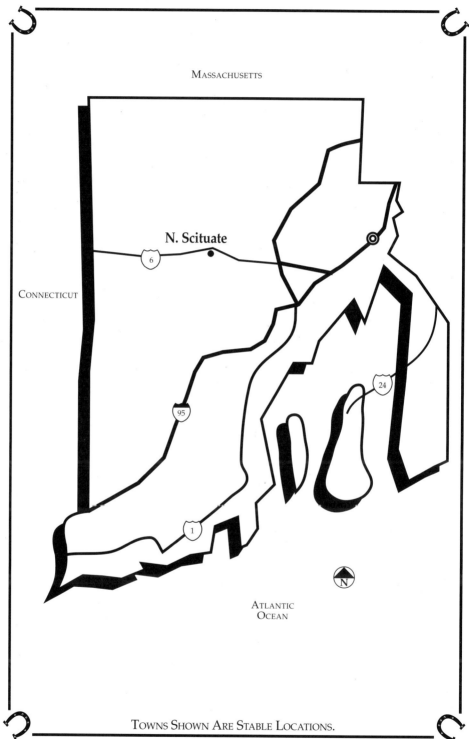

Towns Shown Are Stable Locations.

* onsite accommodations

RHODE ISLAND Page 197

ALL OF OUR STABLES REQUIRE CURRENT NEG. COGGINS, CURRENT HEALTH PAPERS, & OWNERSHIP PAPERS.

NORTH SCITUATE
Stone House Farm Phone: 401-934-0272
George Bessette
86 Peeptoad Road [02857] **Directions:** 15 minutes from Providence off Rt. 6. Call for directions. **Facilities:** 17 indoor stalls, 120' x 80' indoor arena, pasture, and 6 paddocks each over 1 acre. Instruction in hunter/jumper for horses & riders from beginner to Grand Prix. Also buys and sells quality horses on consignment. Call for reservation. **Rates:** $20 per night. **Accommodations:** Motels 5 miles from stable.

PAGE 198 SOUTH CAROLINA

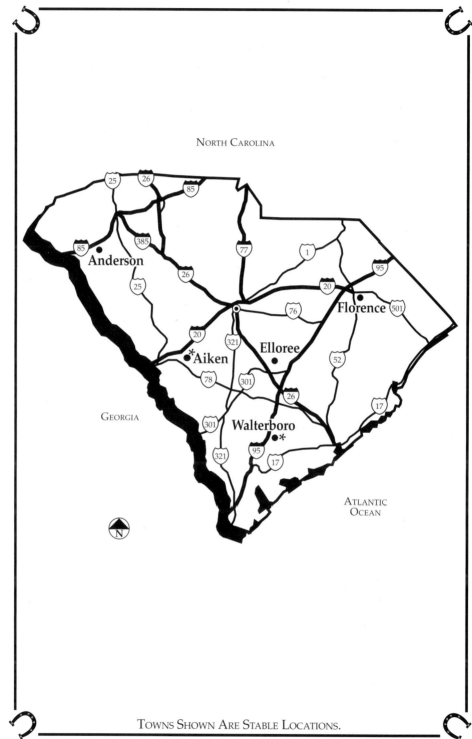

Towns Shown Are Stable Locations.

* onsite accommodations

SOUTH CAROLINA

ALL OF OUR STABLES REQUIRE CURRENT NEG. COGGINS, CURRENT HEALTH PAPERS, & OWNERSHIP PAPERS.

AIKEN

✶ Fulmer International
Robert N. Hall, D.B.H.S.
2500 Dibble Road [29801-3381]
Phone: 803-649-0505 Fax: 803-649-1200
E-mail: fulmer@scescape.net
Web: www.equitation.com
Directions: Aiken-Augusta Hwy 1/78, go South on 118, east on Dibble Road. Go 1/2 mile, just past the natural gas sub-station and turn right up the sandy driveway & continue up to the stable. **Facilities:** 10 indoor stalls, 1 large paddock. Trailer parking available for guests. Please bring your own feed/hay. Stable adjoins the famous 2,000-acre Hitchcock Woods, which has over 40 miles of trails and jumps (hunt fences). Maps available. Boarding facilities, breaking and training through Gran Prix level, student instruction in Dressage, Eventing and Show Jumping. Breeding facilities and home-bred horses for sale. Telephone in advance for reservation. **Rates:** $20 per night. **Accommodations:** Variety available, including B&B.

ANDERSON

Penn's Woods Stable
Bill Payne
Phone: 864-261-8476
1930 Denver Road [29625] **Directions:** Exit 19B off of I-85. Call for directions. **Facilities:** 12 box stalls, 4 tie stalls, pasture, 4 turnout paddocks, outdoor riding ring, jumper & cross-country course & fox hunting. Training of hunter/jumper for horses & riders. Complete blacksmith shop on premises. Near Clemson University. Call for reservation. **Rates:** $20 per night. **Accommodations:** Motels less than 2 miles from stable.

ELLOREE

R.V. Shirer Thoroughbreds
R.V. Shirer, Owner
Phone: 803-897-2238
East Harlin Street (Mail: 6735 Old #6 Hwy.) [29047] **Directions:** From I-95: Take Exit 98 towards Elloree & go 7 miles; 2 streets past only traffic light, turn right. Stable is on right side. **Facilities:** 8 indoor stalls, 5 outdoor stalls, 5 paddocks, hot walker, feed/hay & trailer parking. Campers OK if self-contained. **Rates:** $15 per night; $75 per week. **Accommodations:** Motels in Santee, 7 miles from stable.

FLORENCE

Florence Horse Center
Jack Belew
Web: www.easternx.com
Phone: 843-679-5502 or: 843-667-0951
E-mail: jbelew@sc.rr.com
Fax: 843-667-3504
3508 Cherrywood Road [29501] **Directions:** 5 minutes from I-95 and US 52. Call for directions. **Facilities:** 40 indoor stalls, ring, paddocks, dressage ring, and jumps. Western & English dressage & hunter training for horses and riders. Horses for sale. Truck & trailer parking available. Also, home of Eastern Equine Express Horse Transportation, serving all of the lower 48 states. Reservation required. **Rates:** $25 per night. **Accommodations:** All major motels & restaurants within 5 minutes of barn. See listing on www.easternx.com.

SOUTH CAROLINA

ALL OF OUR STABLES REQUIRE CURRENT NEG. COGGINS, CURRENT HEALTH PAPERS, & OWNERSHIP PAPERS.

WALTERBORO
Double D Stable   Phone: 843-893-3894
Tommie Derry
1256 Rodeo Drive. [29488]Directions: Call for Directions. **Facilities:** 16 12X12 indoor stalls, 4 multi-acre pastures, 2 barns with 8 stalls each, hook-ups for overnight, hot walker, round pen, arena with lights and bleachers, stalls have auto water and overhead lights, wash room with hot and cold water, small pets must be on leash and contained, prefer reservations. **Rates:** $25 per night, $150 per week. **Accommodations:** Holiday Inn, Econo Lodge and many other major hotels nearby.

✱ **Mt. Carmel Farm Bed & Breakfast**   Phone: 843-538-5770
Maureen Macknee
Rt. 2, Box 580A [29488] Directions: 3-1/2 miles off I-95. Reservations required. Call for directions. **Facilities:** 8 stalls, paddocks, turnout, layups, round pen. Trailer parking on site. This is a bed & breakfast for people traveling with horses and/or other animals. Overnight boarding is only for guests of the B & B. **Rates:** Call for reservations. **Accommodations:** B & B with 2 guest rooms with private baths.

Walterboro

Double D Stable

- 16 12X12 indoor stalls
- 4 multi-acre pastures
- two barns with 8 stalls each
- hook-ups for overnight
- hot walker
- round pen
- 175' x 250' arena with lights and bleachers
- stalls have auto water & overhead lights
- wash room with hot & cold water

Note: small pets must be on leash and contained, prefer reservations. **Rates:** $25 per night, $150 per week. Holiday Inn, Econo Lodge and many other major Hotels nearby.

Tommie Derry
1256 Rodeo Drive. Walterboro, SC 29488
Phone: 843-893-3894

PAGE 202 **SOUTH DAKOTA**

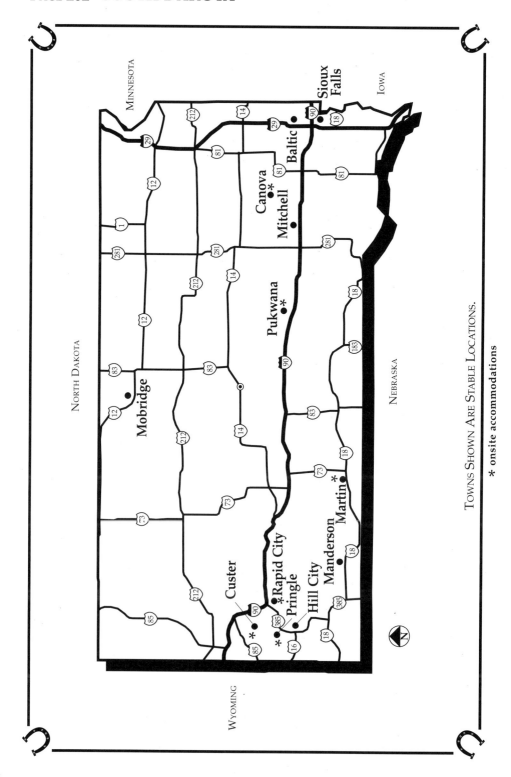

SOUTH DAKOTA

ALL OF OUR STABLES REQUIRE CURRENT NEG. COGGINS, CURRENT HEALTH PAPERS, & OWNERSHIP PAPERS.

BALTIC
Heartland Arabian Farm   Phone: 605-543-5900
Jane & Lloyd Solberg, owners
25467 473rd Avenue [57003] **Directions:** Exit 399 at I-29 & 1-90 junction. Call for directions. **Facilities:** 64 indoor heated 10' x 12' stalls, 3 outside paddocks on a total of 190 acres. Feed/hay & trailer parking available. An Arabian breeding & training farm. 5 stallions standing-at-stud including "The Chief Justice." Horses for sale. Call for reservation. **Rates:** $20 per night; $100 per week. **Accommodations:** Ramkota Inn in N. Sioux Falls, 4 miles from stable.

CANOVA
✱ **Skoglund Farm Bed & Breakfast**   Phone: 605-247-3445
Alden Skoglund
Rt. 1, Box 45 [57321] **Directions:** Call for directions. **Facilities:** 3 outdoor pens, pasture/turnout area, no feed/hay, overnight horse trailer parking available. Boarding only for guests at B & B. **Rates:** No charge. **Accommodations:** Bed & Breakfast on premises: Adults $30, teens $20, children $15, 5 & under, free.

CUSTER
✱ **Spirit Horse Escape**   Phone: 605-673-6005
Barney & Linda Fleming   Fax: 801-730-1481
Web: www.spirithorseescape.com   E-Mail: fleming@gwtc.net
11596 w. Hwy 16 [57730] **Directions:** 7 miles west of Custer SD on Hwy 16 on North side of Hwy. **Facilities:** 6 full hookups, 10 non-hookup sites, outdoor pipe corrals for 25 horses. Trailer parking and round pens available. Many miles of trails on the beautiful Black Hill of SD **Rates:** $10 per night. **Accommodations:** Deluxe rental cabins on property. Many hotels in Custer.

Tex's Horse Motel   Phone: 605-673-5186
Ken Irwin   E-mail: tex@gwtc.net
25321 Rodeo Drive [57730] **Directions:** 3.5 miles west of Custer on Hwy 16. North side of Rd, look for big red barn. **Facilities:** 16 12'x12' horse safe metal pens, 50'x 50' corral. Feed/hay and trailer parking available. **Rates:** $15 per night. **Accommodations:** B&B, motels and cabins all close by.

HILL CITY
Happy Hill Ranch & Tack Shop   Phone: 605-574-2326
Joyce Floyd
23845 Penalua Gulch Road [57745] **Directions:** 3 miles east of Hill City; 1/8 mile off Rtes. 385 & 16. **Facilities:** 8 - 10' x 10' and 10' x 12' stalls, 40' x 60' grassy paddocks, feed/hay, trailer parking. Black Trakehner standing-at-stud; Warm Blood Crosses for sale. Tack shop, training. **Rates:** $10 per night; $60 per week. **Accommodations:** Motels within 3 miles of stable.

SOUTH DAKOTA

ALL OF OUR STABLES REQUIRE CURRENT NEG. COGGINS, CURRENT HEALTH PAPERS, & OWNERSHIP PAPERS.

MANDERSON
Ecoffey Stables   Phone: 605-867-5698
Gilbert Ecoffey
Box 345 [57756] **Directions:** Call for directions. Short distance from Wounded Knee. **Facilities:** 8 indoor stalls, 100' x 150' arena, feed/hay & trailer parking available. Can assist anyone in area if they have broken down. Will get horses & trailer. Call for reservation. **Rates:** $15 per night. **Accommodations:** 3 motels in Gordon, NE 30 miles from stable.

MARTIN
T - (T Bar)   Phone: 605-685-6900
✱ Thomas H. Loomis
HC #2, Box 5A [57551] **Directions:** From I-90: South on 73 to Rt. 18. Straight at stop onto "Old 18" 3.5 miles heading west past Deadman's Lake on south side of road to top of hill turn south, 2 miles on dirt road then east 1/4 mile. **Facilities:** 6 - 7' x 10' box stalls plus corral, 26+ acres of pasture/turnout, feed/hay available, trailer parking. Horse owners care & maintain their own animals. Trail maps for riding available. A confirmed reservation is mandatory! **Rates:** $12 per night; $70 per week. **Accommodations:** 6-bed bunkhouse with 2 showers and outside shower on premises; $35 per person per night.

MITCHELL
Mitchell Livestock Auction Co.   Phone: 605-996-6543
Tim Moody
P.O. Box 516 [57301] **Directions:** Exit 332 off of I-90: go 1/4 mile south on Hwy 37. Look for big sign & go 1/4 mile east. **Facilities:** 200 indoor & outdoor pens. Feed/hay & trailer parking available. **Rates:** $3 per night. **Accommodations:** Super 8 & Best Western 1/4 mile away.

PRINGLE
Plenty Star Ranch   Phone: 605-673-3012
✱ E-mail: plenty@rapidnet.com   Web: www.plentystarranch.com
PO Box 106 [57773] **Directions:** We are on Hwy 385, about 5o miles South of Rapid City, I-90. Call for Directions. **Facilities:** 22 covered outdoor 10'x30' and 12'x24' stalls, feed/hay, trailer parking with electric and water for $15 per night, 3 40'x50' pens, breeding and training or registered Spanish Mustangs, Standing STR Stallions, plus rare Sorraia Mustangs. Marked loop trails leave directly from the camp into the Black Hills National Forest, Ranch restaurant open for breakfast and dinner, family style by reservation. **Rates:** $10 per night. **Accommodations:** We have cabins, tipis, tent and RV sites. Nearest Motel is 9.5 miles away in Custer.

SOUTH DAKOTA PAGE 205

ALL OF OUR STABLES REQUIRE CURRENT NEG. COGGINS, CURRENT HEALTH PAPERS, & OWNERSHIP PAPERS.

PUKWANA
✱ Diamond A Cattle Co **Phone:** 605-778-6885
Crystal & Tucker Ashley   605-730-1074
35540 250th St [57370] **Directions:** Kimball Exit off of I-90: Go 6 miles west on paved road; 1 mile north on gravel, & 3/4 mile west to indoor arena. **Facilities:** 2-10 large pens to turnout, can make other pens with portable panels, large pasture areas, indoor & outdoor roping arenas, feed/hay & trailer parking. Stock available for roping practice. **Rates:** $10 per night. **Accommodations:** Cabin available to rent.

RAPID CITY
✱ Bunkhouse Bed & Breakfast & Working Ranch **Toll Free:** 1-888-756-5462
Carol Hendrickson
Web: www.bbonline.com/sd/bunkhouse
14630 Lower Spring Creek Road, Hermosa [57744] **Directions:** Exit 60 off of I-90 into Rapid City; left at first light onto Cambell St.; go thru 1 more light; at second light, read odometer and go 8 miles south on Hwy 79 to Lower Spring Creek Rd; take left and stable is 4.3 miles. **Facilities:** 3 indoor 10' x 12' & 14' x 14' stalls, 5 outdoor pens with wind break, no pasture/turnout, feed/hay available at $3 per horse, trailer parking available. No stallions. Riding trails on ranch plus trail maps of great trails in area. B & B not open Jan 1 - May 1. **Rates:** $7 or $10 ($3 for hay) per horse per night if guest at B & B; $20 or $23 ($3 for hay) per horse per night if not. **Accommodations:** Bed & Breakfast on premises with 3 guest rooms. Full breakfast, great accommodations. Motels nearby in Rapid City.

SIOUX FALLS
Cedar Ridge Equestrian Center **Phone:** 605-543-5120
Laura Wagner   **Barn:** 605-543-5100
25670 475th Ave [57055] **Directions:** 4 miles North of I-90 Exit 399 on East side of road. Please call ahead. **Facilities:** 30 indoor 10'x10' stalls, 9 pens 30'x40'. Trailer parking available, 80 acres of pasture, new indoor riding arena, 20 plus years experience, negative coggins, current health certificate. **Rates:** $15 per pen, $20 per stall. **Accommodations:** Super 8, Comfort Inn, Cloud 9, Days Inn all in Sioux Falls. KOA campground 4 miles South.

TENNESSEE

Towns Shown Are Stable Locations.

* onsite accommodations

TENNESSEE Page 207

ALL OF OUR STABLES REQUIRE CURRENT NEG. COGGINS, CURRENT HEALTH PAPERS, & OWNERSHIP PAPERS.

BON AQUA
Harmony Farms Phone: 931-670-4737
Joann & Randy Jackowski or: 931-670-6560
10726 Harmony Farm Lane [37025] **Directions:** Exit 172 off of I-40. Go left 4 miles - Harmony Farms on left. **Facilities:** 10+ box stalls, 10+ slip stalls, 40' & 60' round pens, feed store & saddle shop on premises. Trailer parking on premises, up to & including, tractor trailers. Mares with foals & stallions welcome. Thoroughbred Sporthorse breeding program. Standing-at-stud: "EVENING CZAR," "MAKE IT ALL," "SEBASTIAN," "JESTCINO." **Rates:** $15 for box stall, $10 for slip stall per night; $95 per week. **Accommodations:** Comfort Inn & Days Inn nearby.

BRENTWOOD
✶ **Brentwood Bed & Breakfast Inn** Phone: 800-332-4640
Lisa Rosche or: 615-373-4627
6304 Murray Lane [37027] **Directions:** 10 miles south of Nashville off I-65 South. **Facilities:** 3 stalls, 5 acres of pasture, 40' barn and tack room, feed/hay, trailer parking. Call for reservations. **Rates:** Call for rate. **Accommodations:** Upscale B&B with 6 bedrooms with private baths, full breakfast. No pets. A unique destination.

BUTLER
Iron Mountain Inn Phone: 423-768-2446
Vikki Woods Web: www.ironmountaininn.com
138 Moreland Drive [37640] **Directions:** Call for directions. **Facilities:** Two stalls on premises with additional nearby, 10'x12' run-in-sheds, various pasture/turnout areas. Feed/hay available, trailer parking. Located near hundreds of miles of riding trails in Tennessee, Virginia and North Carolina. **Rates:** $15 per night, $60 per week. **Accommodations:** Iron Mountain Inn, which is a B & B that encourages its guests to enjoy the trails, andthe Inn.

CAMDEN
✶ **Bird Song Trail Ride** Phone: 731-584-9206
Pat & Norman Fowler or: 731-584-4280
565 Little Birdsong Road [38320] **Directions:** 85 miles west of Nashville. Exit 126 off I-40, 8 miles north towards Camden, right on Shiloh Church Road for 1 mile, right on Little Birdsong Road. Stable is first place on right. **Facilities:** 300 covered boxed stalls, feed/hay, trailer parking, camper hook-ups. Riding trails along creeks and the Tennessee River; trail rides, ride alone or in groups. **Rates:** $15 per night, weekly rate on request. **Accommodations:** Campsites on premises, electric available. Colonial Inn, Guest House Inn. Brochures available on trail riding and dates.

TENNESSEE

ALL OF OUR STABLES REQUIRE CURRENT NEG. COGGINS, CURRENT HEALTH PAPERS, & OWNERSHIP PAPERS.

CHRISTIANA
Kanawha Farm   Phone: 615-895-9262
Janet E. Stevens   E-mail: kanawha3@aol.com
9996 Manchester Hwy [37037] **Directions:** 1 mile off of I-24 at Exit 89 on Hwy 41E. Call for further directions. **Facilities:** 3 indoor 10' x 10' stalls . One 1/2-acre turnouts with run-ins, feed/hay, trailer parking. Standing-at-stud: black tobiano "Dam Straight", by "Dam Yankee", sport type horses. **Rates:** $20 per night. Require 2 week reservations, coggins & health papers **Accommodations:** Holiday Inn, Best Western, Howard Johnson in Murfreesboro, 9 miles away.

CLEVELAND
BJ's Stables   Phone: 423-339-3783
B. J. Owens   Dan DeFriese, manager
655 Urbana Road NE [37312] **Directions:** Just off I-75, Exit 27. 5 minutes from downtown Cleveland. Call for further directions. **Facilities:** 22 - 12' x 12' stalls, 30 acres of pasture/turnout, feed/hay available, parking area large enough for tractor-trailers. Two wash racks, hot & cold water, lighted round pen and arena. Large animal veterinarian clinic on premises with 24-hour call. **Rates:** $20 per night; $100 per week. **Accommodations:** Holiday Inn, Red Carpet Inn, Scottish Inn, and others 5 minutes from stable.

COLLEGE GROVE
✱ **Peacock Hill Country Inn**   Phone: 615-368-7727
Anita & Walter Ogilvie   or: 800-327-6663
6994 Giles Hill Road [37046] **Directions:** Call for directions. **Facilities:** 8 new indoor 12' x 12' stalls, pasture/turnouts, feed/hay, trailer parking. 650-acre working cattle farm, new barn, miles of trails for riding, hiking. Reservations required. **Rates:** $12 per night. **Accommodations:** Luxury Country Inn on premises, $95-$125 per night includes breakfast.

FAIRVIEW
Best Little Horse House   Phone: 615-799-8833
Gennette S. Norman   Cell: 615-500-8812
7201 Cumberland Dr [37062] **Directions:** 5 miles from I-40, exit 182. 25 miles SW of Nashville. **Facilities:** 6 indoor stall, 2 holding pens and pasture. Close to fishing, camping, hiking, horse trails and Nashville attractions. (Formerly Sweet Annies).

Lazy Susan Overnight Boarding   Phone: 615-799-0991
Rick & Susan Morrison   E-mail: lazysusanrick@bellsouth.net
7250 Northwest Hwy (37062) **Directions:** 3 miles off I-40W, exit 182, 20 miles west of Nashville. **Facilities:** 6 indoor stalls, 60' round pen with 6 acres of pasture. Feed/hay and oats available. **Rates:** Call for rates. **Accommodations:** Many hotels in Nashville.

ALL OF OUR STABLES REQUIRE CURRENT NEG. COGGINS, CURRENT HEALTH PAPERS, & OWNERSHIP PAPERS.

FRANKLIN

* **Namaste Country Ranch Inn** — **Phone: 615-791-0333**
Lisa Winters — **E-mail: namastebb@aol.com**
5436 Leipers Creek Road [37064] — **Web: namastacres.com**
Directions: Call for directions. **Facilities:** Arena, round pen, walker. Quiet valley setting, 26 miles of scenic horse trails, swimming pool & hot tub. Open year round. AAA approved. Horse owners must be guests of B & B. 1 mile off scenic Natchez Trace Parkway, 11 miles from Franklin. **Rates:** $10 stalls. **Accommodations:** Country home offers 3 private suites, in-room coffee, phone, fridge, TV/VCR, firelplace, private entrance and bath. $75-$85.

KNOXVILLE

Cumberland Springs Ranch — **Phone: 865-584-5857**
or: 865-558-0914
Gene French
4102 Sullivan Road [37921] Directions: From I-640: Take Western Ave. Exit (West); turn right on Sullivan Rd.; go about 1 mile & turn right at natural wood fence. **Facilities:** 19 indoor 12' x 20' & 12' x 14' stalls, 2 large outdoor arenas, 60' x 120' indoor arena, grass, hay & feed available. Trailer parking. Also offers local and long-haul horse transportation. "Your horse's safety and comfort is our goal." **Rates:** $25 per night. **Accommodations:** Several top-name motels less than 2 miles away.

Hunter Valley Farm — **Phone: 865-690-6661**
Becky Elmore
9111 Hunter Valley Lane [37922] Directions: I-40 West to Maloney Hood Exit: Go left & go to first light (Kingston Pike) & turn right; at next light turn left (Pellessippi Parkway); go to Northshore Drive Exit & turn left at bottom of ramp; go 1/2 mile to Keller Bend; turn right & go .1 mile & turn left on Hunter Valley Lane. Farm is .1 mile on left. **Facilities:** 10 indoor stalls, large paddocks, feed/hay & trailer parking. **Rates:** $25 per night. **Accommodations:** Red Roof Inn, LaQuinta, & others 10 minutes away.

LEBANON

Cedars of Lebanon Stables — **Phone: 615-444-5465**
Lena Veasey
Cedar Forest Drive [37090] Directions: From I-40, Exit 238 to Hwy 231 south, 6.5 mi. south of Lebanon to Cedars of Lebanon Park. Follow signs to stables. **Facilities:** 19 indoor stalls, 10 acres of pasture/turnout, feed/hay available, trailer parking. Overnight rides, plus hayrides; 1-hour, 2-hour, and half-day rides. **Rates:** $15 per night; $75 per week. **Accommodations:** Best Western in Lebanon, 6 miles from stables.

TENNESSEE

ALL OF OUR STABLES REQUIRE CURRENT NEG, COGGINS, CURRENT HEALTH PAPERS, & OWNERSHIP PAPERS.

LEBANON
Cool Breeze Ranch Phone: 615-443-0347
J.R. & Juli Kelley   or: 615-812-5869   Mobile: 615-969-5498
1400 Peyton Road [37087] **Directions:** Located 1.25 miles off of I-40. Exit 239 B off of I-40 East, Exit 239 off of I-40 West. Turn right on Peyton Rd 1/4 mile from I-40. Ranch is on right. **Facilities:** 17 indoor stalls, 75' x 125' round pen, 2-acre pasture, feed/hay & trailer parking available. **Rates:** stalls-$20 per night; 75'x150' corral-$10/horse/night. **Accommodations:** Eight motels 2 miles from stable.

LYNNVILLE
Rally Point Farm   Phone: 931-527-3800
Ken & Bitsy Latta
250 Yokley Road [38472] **Directions:** 9 miles from I-65, Please Call for Directions. **Facilities:** 8 indoor stalls, feed/hay available, trailer parking, 2 turnout (1/2 acre and 200'), 2 bedroom, 1 bath house with kitchen available. **Rates:** $25 per night; $100 per week. **Accommodations:** Richland Inn, Pulaski (12 miles) Richland Inn, Columbia (17 miles).

MANCHESTER
✷ **The Barn**   Phone: 800-292-5807
Bob Kraft   Or: 931-857-3860
287 Matlock Rd [37355] **Directions:** Off I-24. Call for directions.
Facilities: 9 12x12 indoor stalls. Feed/hay available. Trailer parking available, 5 acre pasture and turnout. **Rates:** $20-25 per night. **Accommodations:** Have apartment on property.

MEMPHIS
The Shelby Farms Show Place Arena   Phone: 901-756-7433
(see ad on opposite page)   Fax: 901-756-9920
Shelby County Government
105 S. Germantown Road [38018] **Directions:** Take Exit 13 East (Walnut Grove Rd.) off of I-240 or take Exit 16 South (Germantown Rd.) off of I-40. Arena located at Walnut Grove & Germantown. **Facilities:** 632 covered 10' x 10' stalls, 6' high cattle holding pen, overnight trailer parking & RV hook-ups on grounds (electric & water only - dump station on grounds) no feed/hay. **Rates:** $17 per night, includes 1 bag shavings. **RESERVATIONS REQUIRED at least 2 weeks in advance.** No security on grounds. **Accommodations:** Hampton Inn (901-747-3700).

TENNESSEE PAGE 211

SHELBY FARMS SHOW PLACE ARENA OF MEMPHIS
105 Germantown Road South Memphis, Tennessee 38018

(901) 756-7433

Your Overnight Stables in the Mid-South

- Central location!
- 75% of US Population lives within a 600-mile radius.
- Over 600 stalls available.
- RESERVATIONS REQUIRED.
- Bedding available for purchase on site.
- RV/Trailer hook-ups available on site.
- All barns equipped with complete fly spray system.
- Full service restaurant located right next door.
- Host hotels within 5 miles.

IMPORTANT: Original 12-month coggins test is required before entry onto Show Place grounds. No exceptions.

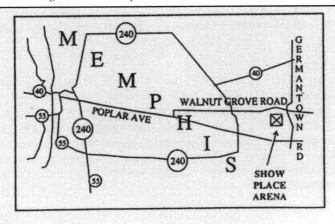

PAGE 212 **TENNESSEE**

ALL OF OUR STABLES REQUIRE CURRENT NEG, COGGINS, CURRENT HEALTH PAPERS, & OWNERSHIP PAPERS.

MURFREESBORO
Womack Stables Phone: 615-896-2310
Ricky Womack Fax: 615-848-9290
Web: glenoaksfarm.com E-mail: womackst@bellsouth.net
4024 Barfield Cresent Rd [37218]Directions: From 24 E exit 81A turn right, 24 W exit 81 turn left. Exit 81 toward Hwy 231 south toward Shelbyville go to 5th light, Barfield Cresent Rd turn right, stable 3 miles on right. **Facilities:** 100 12x12 stalls. Pasture and turnout available. Feed/hay and trailer parking available. Breeding farm of Tennessee Walkers, many at stud. We offer riding lessons by a current world champion rider. **Rates:** $10 per night. **Accommodations:** Shoney's Inn, Quality Inn all 7 miles from stable on Church St.

NASHVILLE
✱ **Apple Brook Bed, Breakfast, and Barn** Phone: 615-646-5082
Donald & Cynthia Van Ryen
9127 Hwy 100 [37221-4502] **Directions:** Exit 199 off I-40 West, south to Hwy 100, right 6.5 miles, stable on left. **Facilities:** 4 indoor stalls, feed/hay on request, trailer parking. Nearby attractions include Natchez Trace, Fairview Nature Park, and many others. **Rates:** $10 per night. **Accommodations:** B&B on premises, 4 rooms, 2 w/private baths.

NIOTA/SWEETWATER
Sweetwater Equestrian Center, Inc. Phone: 800-662-4042
Grady V. & Charlotte J. Maraman or: 423-337-2674
E-mail: charlottemaraman@yahoo.com
1065 County Road 316 [37826] **Directions:** Exit 60 off of I-75: Go north on Hwy 68 for 3 miles; turn left on McMinn County Rd. 316; located approx. 1/2 mile on right. **Facilities:** 20 indoor & 8 outdoor stalls under roof, one acre of pasture/turnout, feed/hay & parking for trailer. Electric hook-up for self-contained campers. **Rates:** $20 per night. **Accommodations:** Days Inn, Comfort Inn, Quality Inn & others located at Exit 60, 3.5 miles from Center.

PINEY FLATS
✱ **Walnut Creek Farm** Phone: 423-538-8931
Jane and Phil Elsea, D.V.M. Fax: 423-915-0371
2490 Enterprise Road [37686] **Directions:** 9 miles south of I-81, scenic upper northeast Tennessee area. Call for detailed directions. **Facilities:** 4 indoor 12' x 12' box stalls, 2 indoor 12' x 24' box stalls (could be divided), several turn-out areas/pastures, feed/hay & trailer parking available. Riding ring, draft horse shoeing stocks, walker, video camera monitoring system in two box stalls, Owner is equine veterinarian. Clinic located close by in Johnson City. **Rates:** $15 per night. **Accommodations:** Two bedroom apartment w/ kitchen and bath adjoining barn or numerous motels in Johnson City and Bristol, each about 10 miles from farm.

TENNESSEE Page 213

ALL OF OUR STABLES REQUIRE CURRENT NEG. COGGINS, CURRENT HEALTH PAPERS, & OWNERSHIP PAPERS.

ROCKFORD
✸ **Porter Brakebill Farm**　　　　　　　　　　　Phone: 865-982-0200
Don Brakebill
803 Martin Mill [37853] **Directions:** Call for directions. **Facilities:** 6 indoor 12' x 12' stalls, 1/2 to 5 acres of pasture/turnout, riding area, trails, round riding pen, feed/hay available, trailer parking. **Rates:** $15 per night. **Accommodations:** Bed & Breakfast with 5 rooms available on premises; $90 per night includes "bountiful" continental breakfast. Motels in Maryville and Knoxville, 10-15 miles.

SPRING CREEK
Journey's End　　　　　　　　　　　　　　Phone: 731-697-9678
James & Micki Page　　　　　　　　　　　　　Fax: 731-935-2026
Web: journeysendappaloosa.com　　　　E-mail: pages@charter.net
103 Springcreek-Law Rd Jackson [38305] **Directions:** I-40 Exit 93 Law Rd go north turn left on Hwy 152 for 3.5 miles to 103 Springcreek-Law Rd. Large sign at driveway. **Facilities:** 25 10x15 and 20x24 stalls. Feed/hay available. Trailer parking available with 30-amp hookup. 3 acres of pasture. 24 hour vet service. Negative and health certificate required. **Rates:** $25 per night includes bedding, $140 per week. **Accommodations:** Howard Johnson Exit 85 in Jackson.

TOWNSEND
Davy Crockett Stables　　　　　　　　　　　Phone: 865-448-6411
J.C. Morgan
234 Stables Drive [37882] **Directions:** 30 miles from I-40 and I-75. Call for directions. **Facilities:** 10 indoor stalls, no pasture/turnout, sweet feed/hay available. Trailer parking on premises. No stallions. Stable property joins Great Smoky Mountain National Park. **Rates:** $10 per night.
Accommodations: Tremont Campground 200 yards away. Also eight motels 3 miles from stable.

✸ **Gilbertson's Lazy Horse Retreat**　　　　　Phone: 865-448-6810
Melody H. Gilbertson　　　　　　　　　Web: www.lazyhorseretreat.com
E-mail: lazyhrse@icx.net
938 Schoolhouse Gap Road [37882] **Directions:** Entering Knoxville on I-75, do not take I-75 east exit, but continue south on 275. Take Hwy 129 to Alcoa/Maryville (past airport), then 321 north to townsend. **Facilities:** 12 indoor 10' x 10' stalls with dutch doors, 4 corrals approximately 80' x 100', 1/2-acre pasture, feed available, trailer parking. **Rates:** $10 per night in stalls, $5 per night in corral; weekly rate, $60 in stalls, $30 in corrals. **Accommodations:** Two-bedroom cabins on premises with jacuzzi, fireplace, TV-VCR, central heat & air, equipped kitchen, phone, washer/dryer. Also 1-bedroom efficiency cabin. 1.5 miles from Great Smoky Mountain National Park.

ALL OF OUR STABLES REQUIRE CURRENT NEG. COGGINS, CURRENT HEALTH PAPERS, & OWNERSHIP PAPERS.

TOWNSEND

✱ **Packs Boarding Stables** Phone: 865-448-6318
Greg & Freida Pack Web: packsboardingstables.com
E-mail: freidapack@msn.com
7728 Cedar Creek Rd [37882] **Directions:** 1 mile from US 321. Call for directions. **Facilities:** 24 indoor stalls. Feed/hay supplied. Trailer parking and hookup available. Advance notice required for studs. **Rates:** $12 per horse. **Accommodations:** Bunkhouse and cabins available. Many motels within 3 miles of stable.

WALLAND

✱ **Twin Valley Bed & Breakfast Horse Ranch** Phone: 865-984-0980
Janice Tipton
2848 Old Chilhowee Road [37886] **Directions:** Call for directions.
Facilities: 14 indoor & outdoor covered stalls, 1/2-acre corral with creek & shelter, paddock with shelter, wash rack, pond for swimming & fishing, hiking & riding trails, trailer parking available. 12 miles from Great Smoky Mountain National Park. Must sign a liability release form. Must call for reservation.
Rates: $10 per night including hay. **Accommodations:** Bed & Breakfast & cabins on premises. Ranch offers many activities including riding lessons, fishing, BBQs, etc.

NOTES AND REMINDERS

TEXAS Page 217

ALL OF OUR STABLES REQUIRE CURRENT NEG. COGGINS, CURRENT HEALTH PAPERS, & OWNERSHIP PAPERS.

ABILENE
Expo Center of Taylor County, Inc. Phone: 325-677-4376
Web: taylorcountyexpocenter.com
1700 Hwy 36 [79602] **Directions:** From IH 20, take Loop 322 south to Hwy 36. The Expo Center is located on the corner of Loop 322 and Hwy 36. **Facilities:** 750 indoor stalls, limited pasture/turnout. Trailer parking on the 117-acre facility. Feed stores 2 miles from Center. Reservation required. **Rates:** $12 per night. **Accommodations:** Travelodge & Garden Inn 1-3 miles.

AMARILLO
Amarillo Tri-State Exposition Phone: 806-376-7767
Virgil Bartlett Night Pager: 806-342-8141
E-mail: cbackus@wtxcoxmail.com Web: www.amarillonationalcenter.net
3301 E. 10th [79104] **Directions:** Exit I-40 E (Exit # 72B) to Grand and travel 2 miles to entrance. **Facilities:** Unlimited stalls available, 10x12 & 10x10 new stalls all with asphalt floor, trailer parking, shavings can be purchased at the fairgounds. **Rates:** $12-$15 per night. **Accommodations:** Holiday Inn Holidome-(806) 372-8741,Days Inn-(806) 379-6255, and Kiva Inn & Suites- (806)379-6555 nearby.

EE Arena Phone: 806-379-8866
Lee Blakeney E-mail: eearena@niinet.net
1300 SE 46th Ave [79118] Web: www.equestrianequipper.com
Directions: I-40 South on Ross-Osage. Go 2 miles & turn right. Barn is visible from Osage. Turn in at EE Arena sign at the top of hill on left. **Facilities:** 50 12'x12' box stalls. Alfalfa available, large enough for semis. Round pens, indoor & outdoor arena. **Rates:** $20 per night. **Accommodations:** 2 miles from Holiday Inn, La Quinta, Days Inn, and Motel 6.

Tascosa Stables, Inc. Phone: 806-342-9061
Cheryl Rhoderick Toll Free: 866-272-3996
E-mail: tascosastables@earthlink.net Web: www.tascosastables.com
3513 N. Western [79124] **Directions:** Approximately 4 miles north of I-40 and 1/2 mile south of Loop 335. **Facilities:** 35 12x15 indoor with automatic waste disposal, and organic fly spray system. Dutex doors, outside runs, full arena, 2 round pens, 8 horse walker, feed/hay, trailer parking, security gates. Must have health papers and current Coggins test. **Rates:** $20 per night **Accommodations:** Discount motel reservations arranged for Tascosa Stables customers.

ANNA
* **Ararat Acres** Phone: 972-924-3401
Charlie & Liz James Web: www.araratacres.com
11548 Sheffield Dr. [75409] **Directions:** Please Call or visit www.araratacres.com for directions and more information. **Facilities:** 2 12'x12' stalls, 2 12'x14' stalls with outdoor runs, feed/hay, trailer parking, 3/4 acre turnout, 2 acre turnout, several larger pastures, lighted round pen, quality paint horses for sale, CJF farrier available on call, equine vet available, human massage therapist available on-call. **Rates:** $10 per night (feed & hay extra). **Accommodations:** Bed & Breakfast on site. Several motels available 16 miles away.

PAGE 218 **TEXAS**

<u>ALL OF OUR STABLES REQUIRE CURRENT NEG. COGGINS,
CURRENT HEALTH PAPERS, & OWNERSHIP PAPERS.</u>

BELLVILLE
<u>Banner Farm Bed & Breakfast</u> Phone: 800-865-8534
Toni Trimble
681 Farm Road 331 [77418] **Directions:** Exit 720 off I-10, go 4.5 miles north on Hwy 36, turn right at Farm Road 331, facility is 5.5 miles on right. **Facilities:** 9 indoor 10' x 10' and 10' x 12' stalls, 1/3-acre turnout paddocks, lighted riding arena, hay, lighted trailer parking. Call for reservations. **Rates:** $15 per night. **Accommodations:** B&B on premises in turn-of-century farmhouse, 3 guest rooms with shared bath, swimming pool, $60 per night.

CAT SPRING
✶ <u>Southwind Horse Farm</u> Phone: 979-992-3270
Sunny & John Snyder
Rt. 1, Box 15-C [78933] **Directions:** Just west of Houston & 18 minutes north of I-10. Take Sealy/Hwy 36 North (#720) Exit. Call for detailed directions. **Facilities:** 11 large indoor stalls, large turnout area, 4 holding/exercise pens, pasture, 15 acres for riding. Can safely accommodate mare with foal or stallions. Camper hook-up and vet nearby if needed. Kennel for small dogs. Reservations not required but recommended. **Rates:** $15 per night, discounts for 3 or more head. **Accommodations:** Bed & Breakfast on premises with private baths. Discounted B & B/stable rate available.

CORPUS CHRISTI
<u>Golden Gait Farm,</u> Phone: 713-939-7828
Dana Riley or: 713-939-8330
1616 Ramfield Road [78418] **Directions:** Call for directions. **Facilities:** Four 100' x 40' and one 200' x 40' paddocks, lighted arena, feed/hay & trailer parking available. Hunter/jumper & dressage training. Reservation required. No arrivals after 11 P.M. **Rates:** $15 per night. **Accommodations:** Motel 6 & LaQuinta about 10 minutes away.

DALLAS
<u>Bettina Stable</u> Home Phone: 214-343-4747
Irmgard Christina Pomper Office Phone: 214-361-8300
E-mail: pompbarn@earthlink.net
7921 Goforth Road at White Rock Trail [75238] **Directions:** Located in NE Dallas: From LBJ Freeway go south on Audelia, right on Kingsley, left on White Rock Trail & then right on Goforth Road. Stable is on corner, big white barn with red roof. **Facilities:** Indoor stalls. Would prefer 2 days notice. Located in the city of Dallas. **Rates:** $25 with own feed; short & long term rates negotiable. **Accommodations:** Comfort Inn, Doubletree Hotel, and others & within 5 min.

TEXAS PAGE 219

*ALL OF OUR STABLES REQUIRE CURRENT NEG. COGGINS,
CURRENT HEALTH PAPERS, & OWNERSHIP PAPERS.*

EL PASO
Horseshoe Stables   Phone: 915-877-2182
Paul Billingsley   E-mail: horseshoestable@cs.com
924 Gato Rd [79932] **Directions:** I-10 to Artcraft W. to Westside Dr. N. to Gato W. to stables. **Facilities:** 20 indoor stalls, 30 outdoor pens, feed, 2 outdoor arenas, walker and miles of trails. Discount on herds. **Rates:** Call for rates. **Accommodations:** Many hotels in El Paso.

FORT WORTH - See WEATHERSFORD

HANKAMER
* **T & L Enterprises**   Phone: 409-374-2539
Lee & Trudy Masters
HC 1 Box 581 [77560] **Directions:** 45 minutes east of Houston, 30 minutes West of Beaumont, 3 miles North of I-10. **Facilities:** 4 covered 12x12 stalls, feed/hay, trailer parking, covered round pen & pasture. **Rates:** $15 per night. **Accommodations:** Cabin on premises with kitchen accommodating 4 adults w/ continental breakfast.

HOUSTON
Magic Moments Stable   Phone: 713-461-1228
Granger & Jean Durdin, Owners,   Peter Smit, Manager
1726 Upland Drive [77043] **Directions:** I-10 to Wilcrest exit, Go North, immediate left onto Old Katy(west), immediate right onto Upland Drive(North). Second stop sign, Barn is at Upland and Chatterton. **Facilities:** 12x12 indoor stalls with rubber matted floors, Individual 100x100 paddocks. Trailer parking, feed/hay, 24 hour security, lessons and full board. Mostly Arabians and 1/2 Arabians. Appaloosa stud "Thunder Bay May" with references. Vet and farrier on call. Call ahead for reservations. **Rates:** $15 per night. **Accommodations:** La Quinta, Houston (1 mi).

HOUSTON/ CAT SPRING
Rancho Texcelente IXL   Phone: 979-865-3636
Nancy Flick   Fax: 979-865-1929   E-mail: ixl@paso.net   Web: www.paso.net
14012 Paso Fino Rd. (Mail: P.O. Box 55) FM 1094 [78933] **Directions:** from I-10: 50 miles west of Houston. Take exit 720 & go north on 36 for 2 miles, take left on FM 1094 & go 12 miles. Ranch is on left side. **Facilities:** 39 indoor 12'x12' stalls, 4 holding pens, 250 acres of pasture, feed, 1 open and 1 covered arena, covered round pen, walker, escorted or unescorted trails, camper hookup, one kennel, Local horse transportation available. Advance reservations preferred. **Rates:** $15 per night. **Accommodations:** 3 2-bedroom guest houses. Also, Hotel Wayne in Bellville & Best Western Sealy, about 12 miles from stable.

ALL OF OUR STABLES REQUIRE CURRENT NEG. COGGINS, CURRENT HEALTH PAPERS, & OWNERSHIP PAPERS.

INGRAM
<u>Sweet Ranch</u>   Phone: 830-367-3693
John & Rose Sweet
299 Indian Creek Road [78025] **Directions:** Ranch is 5 miles from I-10. Call for directions. **Facilities:** 8 12'x12' stalls. Feed and hay available, trailer parking available. 2 acres of pasture, round pen and automatic waterers in each stall. **Rates:** $20 per night. **Accommodations:** Days Inn, Motel 6, YO Ranch Resort Hotel, Comfort Inn, Hampton Inn, Super 8, Best Western of the Hills. All motels are 4 miles from ranch in Kerrville.

LAREDO
<u>El Primero Training Center, Inc.</u>   Phone: 956-723-5436
Keith Asmussen   956-722-4532
Box 1861 [78044] **Directions:** I-35 to Hwy 59E. Call for directions. **Facilities:** 392 indoor box stalls, 57 holding pens, round pens, 5/8-mile racetrack, feed/hay & trailer parking available. Call for reservation. **Rates:** $12 per night. **Accommodations:** Motels 1 mile from stable.

LUBBOCK
<u>Four Bar K Ranch</u>   Phone: 806-789-8682
Chuck Kershner
2811 98th Street [79423] **Directions:** I-27 Exit on 98th St. Go west 2 miles. Cross over University. Go four blocks. Gate on left. **Facilities:** 8 indoor & 8 outdoor stalls, large riding pasture, turnout areas, roping arena, round pen, 2 outdoor arenas. Summer horse camp for kids/adults. Private lessons year round. Overnight camping. Neg. coggins papers. Call for reservations. **Rates:** $15 per night, $20 with feed, $75 weekly rate. $20 RV hookup, Free trailer parking. **Accommodations:** Holiday Inn, Lubbock Plaza, 2 miles from stable. Marriott 1.5 miles from stable.s

MAGNOLIA
<u>Whisper Breeze Farm</u>   Phone: 936-273-2398
Denise Jones, John Coleman   Beeper: 281-490-1730
1826 Cattle Drive [77354] **Directions:** Approx. 10 miles west of I-45 on FM 1488. Call for exact directions. **Facilities:** 3 indoor 12' x 12' stalls, 2 indoor 12' x 18' stalls, 3 pastures, 1 paddock for turnout, feed/hay available, trailer parking and RV hookups available. Many miles of riding trails available. Call for reservations. **Rates:** $15 per night; weekly rate negotiable. $15 for RV hook-up. **Accommodations:** Many motels in Conroe, The Woodlands, and Tomball, all within 12 miles of farm.

ALL OF OUR STABLES REQUIRE CURRENT NEG. COGGINS, CURRENT HEALTH PAPERS, & OWNERSHIP PAPERS.

McDADE
Martin Ranch Phone: 512-273-9027
Gary W. Martin
R.R. #1 Box 19 [78650] **Directions:** 9 miles east of Elgin on 290. Turn north on Marlin; 3.7 miles to road's end and ranch. **Facilities:** 33 stalls, 24' x 24' or 2.5 acres of pasture/turnout, coastal alfalfa and grains available, trailer parking. RV hook-up. Riding & training areas available. Standing-at-stud: B&W Overo "Sonny's Creation," Chestnut Overo "Sir Teddy Clue," B&W Tobiano "Stormin King." **Rates:** $14 per night; $70 per week. **Accommodations:** Motels in Elgin, 12 miles from ranch.

McKINNEY
McKinney Stables Phone: 972-562-9302
Terri & Bryan Collins
807 Hwy 380 East [75069] **Directions:** Exactly 2 miles east of Hwy 75 on Hwy 380, next to Mobil station. **Facilities:** 65 indoor oversize stalls, no pasture/turnout, oats/coastal hay, and parking for any size truck & trailer. Visa, MC, Discover, & AmEx welcome. **Rates:** $15 per night; $45 per week. **Accommodations:** Holiday Inn, Comfort Inn, & Super 8 Motel all within 2 miles of stable.

ORANGE GROVE
LUR Ranch Phone: 361-384-9118
E-mail: LTUMFLEET@msn.com
459 Co. Rd 300 [78372] **Directions:** Call for directions. **Facilities:** 6 indoor stalls, 1-100x100, 1-130x250 arenas, 10 miles of trails with camper hook-up available. 16 areas of pasture, feed/hay provided. **Rates:** Call for rates. **Accommodations:** Many motels in area.

SAN ANGELO
San Angelo Horse Center, Inc. Phone: 325-374-5391
Trish L. Hutchinson or: 325-658-6613
1845 W. FM 2105 [76901] **Directions:** Between HWY 87 North & S.H. 208. **Facilities:** 12 12x12 box stalls, 10 16x20 outdoor pens, 3-20 acre turnout: cross fenced, lighted arena, feed/hay, trailer parking (no hookups), secure property with keypad gate entry, private 20 acre riding area. Owners live on premises. 1 min. from Collisseum, 5 min. from State Park equestrian center. Negative Coggina and ownership papers. Reservations. **Rates:** Inside-$25 per night, $150 per week. Outside-$20 per night. $120 per week. **Accommodations:** Inn of the Conchos (2mi), Motel 6 (5mi).

PAGE 222 **TEXAS**

ALL OF OUR STABLES REQUIRE CURRENT NEG. COGGINS, CURRENT HEALTH PAPERS, & OWNERSHIP PAPERS.

SAN ANTONIO
✱ T-Slash-Bar Ranch   Phone: 210-677-0502
Roy Thompson
13901 Hwy 90 West [78245] **Directions:** Entrance is located on US Hwy 90 at state Hwy 211. Call with questions. **Facilities:** At least 5 indoor box stalls, outdoor with runs, outdoor pens, numerous pasture/turnout, feed/hay and trailer parking available. Lighted arena, team roping/penning, full boarding/care, access to vet/farrier. Located near Sea World of Texas, Hyatt Resort, Historic Castroville, Kelly/Lackland AFB. **Rates:** $10 per night; weekly rate negotiable. **Accommodations:** Bed & breakfast available at ranch in two wonderful suites, with full kitchen.

SHERMAN
TLC Quarter Horse Ranch & Equestrian Center   Phone: 817-786-2484
4158 Refuge Road [75092] **Directions:** From I-35 go east on US 82 or from US 75 go west on US 82. North on FM 1417 for 4.5 miles, left on Refuge Road 4 miles. Blue barns on left. **Facilities:** 12 indoor 12' x 12' box stalls, 5 indoor 13' x 20' covered 3-sided shed, 4 outdoor stalls (12' x 12', 12' x 16', 12' x 36', 12' x 24'); 5-40 acres of pasture/turnout, outdoor arena, round pen, indoor working area, horse wash facilities; alfalfa and grass, grains available; trailer parking and RV hook-ups. Horses for sale, all ages and levels of training. **Rates:** $10-$15 per night. **Accommodations:** Comfort Inn, Best Western in Sherman, 12 miles away.

SWEETWATER
Ranch House Motel & Restaurant   Phone: 325-236-6341
Sam Patel   800-822-5361
Web: www.sweetwaterranchhouse.com
301 SW Georgia Avenue [79556] **Directions:** I-20, exit 244, Call for Directions. **Facilities:** 9 outdoor locked and lighted stalls. **Rates:** $12.50 per night. **Accommodations:** Full service Motel & Restaurant with 49 spacious rooms, small pets OK, free full breakfast, stalls available for motel guests only.

TOMBALL
Circle G Stables   Phone: 713-698-3456
Sim & Pat Gounarides   (voice pager)
19406 Lindsey Lane [77375] **Directions:** From I-290 West: Go to Telge Road & go north to Self Road; left on Self to dead-end; asphalt road to the right & stable is first on right. **Facilities:** 11 indoor stalls, 7 paddocks, 12' x 24' birthing stall, feed/hay & trailer parking available. Feed store nearby. Please call ahead. "Come and feel at home." **Rates:** $15 per night; $75 per week. **Accommodations:** Best Western on Rt. 249, 6 miles from stable.

TOYAH
Ingram Ranch   Phone: 432-259-3951
Gary Ingram
Box 15 [79785] **Directions:** Seven Blocks North of I-20 **Facilities:** 12 indoor stalls, feed/hay available, trailer parking, arena. **Rates:** $15 per night. **Accommodations:** Motels available in Pecos, Texas (17 miles East).

TEXAS

ALL OF OUR STABLES REQUIRE CURRENT NEG. COGGINS, CURRENT HEALTH PAPERS, & OWNERSHIP PAPERS.

TYLER
Pine Lake Stables　　　　　　　　　　　Phone: 903-592-8075
Joanne Casmo　　　　　　　　　　　　Web: www.pinelakestables.com
11015 Pine Lake Blvd. [75709] **Directions:** Only 15 miles off of I-20. Call for exact directions. **Facilities:** 34 indoor 12' x 12' stalls, wash rack, walker, outdoor arena, round pen, hay, chips, trailer parking, camper/RV hook-ups available. Reservations preferred, but not necessary. Several studds available for breeding (App, Qtr & Paints) **Rates:** $15 per night, bedding extra. **Accommodations:** $15 per night. Motels within 7 miles in Tyler.

WICHITA FALLS
Cruse Acres (see following page)

Turtle Creek Stables & Arena　　　　　　Phone: 817-692-8130
Tambra Holcomb　　　　　　　　　　　　Barn: 817-691-6291
2110 Turtle Creek [76304] **Directions:** Call for directions. Only facility in Wichita Falls. **Facilities:** 12 indoor 18' x 15' treated pine stalls with 40' private runs, pasture and turnout in various sizes, 2 lighted arenas, 100' round lighted pen, hot walker, feed/hay, trailer parking, RV hook-up. Call for reservations. **Rates:** $15 per night; variable rates for multiple horses. Shavings available. **Accommodations:** Motels within 5 miles of stable.

WILLIS
Diamond T Farms　　　　　　　　　　Phone: 409-856-7709
Ron & Linda Tullis　　　　　　　　　　　　or 800-687-0944
12110 Maggie Lane [77378] **Directions:** From I-45: Take Exit 92 west toward lake; go 5 miles, turn right on Cude Cemetary Rd.; go approx. 1 city block & take right on Maggie Lane; go to end of cul-de-sac; stable is last house on left next to the blue/white barn. **Facilities:** 4 indoor 12' x 12' stalls with auto waterers, 2 pastures, round pen, indoor wash rack, tack room, riding area. Space for self-contained campers. Vet/farrier on call. 1 mile from Lake Conroe. Reservations requested. Home of Diamond T Transportation, equine transportation service. **Rates:** $5 - $15 per night; weekly rate negotiable. **Accommodations:** Ramada Inn & Woodlands Inn 10 miles from stable.

ALL OF OUR STABLES REQUIRE CURRENT NEG. COGGINS, CURRENT HEALTH PAPERS, & OWNERSHIP PAPERS.

WICHITA FALLS

Cruse Acres   Phone: 940-767-9284
Sam Cruse   E-mail: scruse@wf.net
Web: www.cruseacres.com or www.horsemotel.com
2550 Windhorst Road (Mail: P.O. Box 8045) [76307] **Directions:** Overheard Expressway @ US 287/281 (South) interchange- Take Windhorst or Midwestern Parkway exit. **Facilities:** Cruse Acres is a 200 acre clean, secure and conveniently located equine facility within the city of Wichita Falls proper. Minutes away from J.S. Bridwell Agricultural Center/Multi-Purpose Events Center and Coliseum. Covered stalls, shed row, stall mats, steels pens, pasture turn-out, round pen (all well lit), bedding & hay included. **Rates:** $25 per night per horse with reservations suggested. Immediate expressway access and city-country location make this *Equine Inn* a facility without equal in the North Texas area. Owners live on premises with Vet on call. Health certificates and negative Coggins <u>must be presented for all horses.</u> **Accomodations:** Howard Johnson, Express Inn near stable.

CRUSE ACRES
of Wichita Falls
Equine Overnight Boarding

Great Vacation Destination
Near New Gateway Water Park &
J.S. Bridwell Agriculteral Center/
Multi-Purpose Events Center & Coliseum

Cruse Acres offers: covered stalls, shed row, stall mats, steels pens, pasture turn-out, round pen (all well lit), with bedding & hay.

940-767-9284

2550 Windhorst Road (Mail: P.O. Box 8045), Wichita Falls, TX 76307

TEXAS PAGE 225

ALL OF OUR STABLES REQUIRE CURRENT NEG. COGGINS, CURRENT HEALTH PAPERS, & OWNERSHIP PAPERS.

WEATHERFORD/FORT WORTH

* **Hidden Lakes Riding Stable**　　　　　Phone: 800-935-0397
Chris Willingham　　　　　　　　　　　　or: 817-448-9910
5400 White Settlement Road [76087] **Directions:** 5 minutes from intersection of I-20, I-30, & FM 1187. Call for directions. **Facilities:** 31 indoor 15' x 15' insulated stalls, 30 acres of pasture/turnout, feed/hay & trailer parking available. **Rates:** $15 per night; $65 per week. **Accommodations:** Two small furnished apartments $20. One suite $35 per night.

HIDDEN LAKES STABLE

- 15 Minutes from Will Rogers Arena
- 4 Miles North of Intersection 1-30/120 & 3325/1187
- 31 15' X 15' Insulated Stalls
- Fly Spray System
- Automatic Waterers
- 2 Arenas 250' X 125', 75' X 100'
- Furnished Suites $25.00 Per Night
- Very Private
- Secure Lot for Rigs
- Commercial Rates for Commercial Haulers
- MUST HAVE CURRENT COGGINS
- RESERVATIONS REQUIRED

5400 WHITE SETTLEMENT ROAD
WEATHERFORD, TX 76087
817-448-9910

Page 226 UTAH

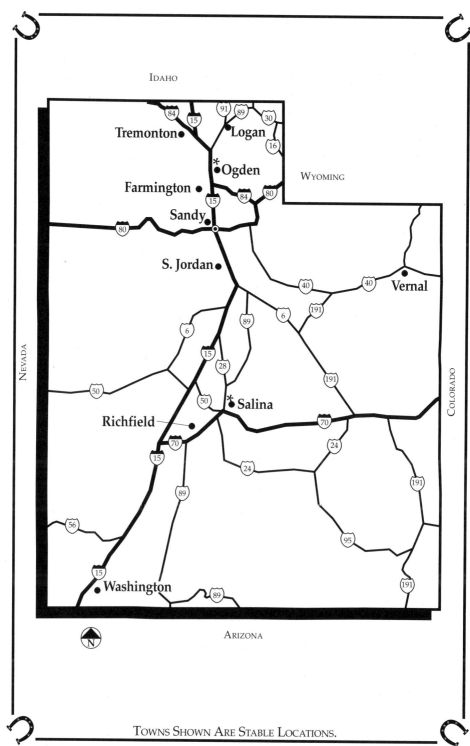

Towns Shown Are Stable Locations.

* onsite accommodations

UTAH Page 227

ALL OF OUR STABLES REQUIRE CURRENT NEG. COGGINS, CURRENT HEALTH PAPERS, & OWNERSHIP PAPERS.

FARMINGTON
Davis County Fair Park & Legacy Center Phone: 801-451-4080
Web: www.daviscountyutah.gov/fairpark Fax: 801-451-4081
E-mail: ckoch@co.davis.ut.us
151 S 111 W [84041] Directions: South bound traffic on I-15 take exit 327, follow signs. North bound I-15 take exit 326, Park/Lane/Farmington, then exit 334 Parklane and follow the signs. **Facilities:** 166 covered stalls, 134 uncovered stalls, indoor arena with open ride time, and round pen. **Rates:** $15 first night, $10 each night after. Shavings provided. **Accommodations:** Several hotels within a 10 mile radius. Call for information.

LOGAN
Logan City/ Cache County Fairgrounds Office Phone: 435-750-9896
Michael Twitchell, manager Cell Phone: 435-757-8574
400 South 500 West [84321] Directions: 21 miles from I-15 and I-84, 100 miles from Salt Lake City. Call for Directions. **Facilities:** 60 12'x12' indoor covered stalls, trailer parking, 225x125 indoor arena, 250x150 outdoor arena, 2 exercise arenas, cross-country jumping course, 1/2 mile race track, Overnight camping on grass next to stream with electric hookup. this is a rodeo facility on 20 acres hosting big school college and PRC rodeos. **Rates:** $10 per night. **Accommodations:** Days Inn, Super 8, and Best Western in Logan within 5 blocks.

OGDEN
Golden Spike Event Center Phone: 800-44-ARENA
Weber County Fairgrounds
Web: www.goldenspikeeventcenter.com E-mail: scall@co.weber.ut.us
1000 N.1200 West [84404] Directions: Take Exit 346 (Harrisville) off I-15 and follow signs to fairgrounds. **Facilities:** 549 covered outdoor 10' x 10' stalls, trailer parking available. Call for reservations. **Rates:** $15 per night. **Accommodations:** High Country Inn, Comfort Inn, Holiday Inn Express, Western Inn, within 2 miles.

Western Inn with Stables Phone: 801-731-6500
✱ Koreann Rael Fax: 801-731-6282
E-mail: koreann@westerninnogden.net Web: www.westerninnogden.com
1155 S. 1700 W. [84404] Directions: I-15 (84) 12th Street exit 347 **Facilities:** 5 stalls, 3-10'x10' 1 20'x20'. Trailer parking available. **Rates:** $10 per night. **Accommodations:** We are a motel.

SALINA
Best Western Shaheen Equestrian Motel Phone: 801-529-7455
✱ Larry Shaheen
1225 S. State Street [84654] Directions: Exit 54 off of I-70. Motel is 1,000 yards north of exit. Located on or near US 50 & US 89. **Facilities:** 61 Horse stalls, trailer parking available. Adjacent to Blackhawk Arena, an equestrian event center. Reservations preferred. **Rates:** $10 per night; $60 per week. **Accommodations:** Best Western hotel and restaurant on premises.

ALL OF OUR STABLES REQUIRE CURRENT NEG. COGGINS, CURRENT HEALTH PAPERS, & OWNERSHIP PAPERS.

SANDY
Alta Hills Farm　　　　　　　　　　　　　　　**Phone:** 801-571-1712
C. Diane Knight
10852 South 2000 East [84092] **Directions:** I-15 to 10600 south, turn east 20 blocks, turn right 1-1/2 blocks south. **Facilities:** 35 indoor stalls, feed/hay, trailer parking. **Rates:** $20 per night. **Accommodations:** Motels nearby.

SOUTH JORDAN
Terry Teeples Horse Boarding　　　　　　　**Phone:** 801-446-8343
Terry Teeples　　　　　　　　　**Web:** www.terryteeplesstables.com
11040 So. 2700 West so. Jordan (84095) **Directions:** I-15 Freeway go west at exit #297 (10600 So.) Go to 2700 west and go left (South) go 3/4 mile to 11040 So. 2700west and go right (west) down dirt lane that dead ends at the barns. **Facilities:** 110 12x12 inside, bigger outside stalls with roof covering. Feed/hay and trailer parking available. Pasture and turnout. **Rates:** $10 per night, $250 a month. **Accommodations:** Super 8 motel at I-15 Freeway exit 297 2 miles from stables.

TREMONTON
Box Elder County Fairgrounds　　　　　　　**Phone:** 435-257-5366
Bill Smoot, Manager
400 N. 1000 West [84337] **Directions:** Take Exit 40 off of I-15. You will see the fairgrounds. **Facilities:** 120 outdoor, completely covered stalls with bedding, large indoor arena, round pen, large open corrals, water accessible. Call 24 hours a day for information. No need to call in advance but it would be helpful. Horse shows, reining & rodeos, year-round team penning held at fairgrounds. **Rates:** $10 w/ bedding for stall, $10 for corrals per night. **Accommodations:** Western Inn next to fairgrounds, Sandman Motel 2 miles away.

VERNAL
Western Park　　　　　　　　　　　　　　　**Phone:** 435-789-7396
Derk Hatch
300 E. 200 S [84078] **Directions:** 2 blocks from Hwy 40. Call for directions. **Facilities:** 400 covered stalls, 102' x 203' indoor arena, 160' x 270' outdoor arena, 5/8-mile racetrack, convention center, Old West Museum, amphitheatre, playground, & trailer parking. **Rates:** $10 per night. **Accommodations:** EconoLodge & Best Western 1 mile from stable.

WASHINGTON
Harmony Horse Haven　　　　　　　　　　**Phone:** 435-673-3991
Steven L. Hafen　　　　　　　　　　　　　　　　or: 435-680-2650
2321 S. Washington Field Rd. [84780] **Directions:** I-15 to exit 10, in Washington to 300 East (only light), turn right or South, go 2 1/4 miles across bridge by church on left, around hill, white fence, pine trees every 25'. 2-1/2 miles from Washington light. **Facilities:** 24 - 12'x36' outdoor stalls w/cover, feed/hay available, plenty of trailer parking, arena. Pretty Wilkin, Standing at stud, foundation quarter horse. V.E.W.T. Flu & Rino shots required. **Rates:** $10/$12 per day, ask for weekly rates. **Accommodations:** Red Cliffs Inn, 3 miles from stable.

NOTES AND REMINDERS

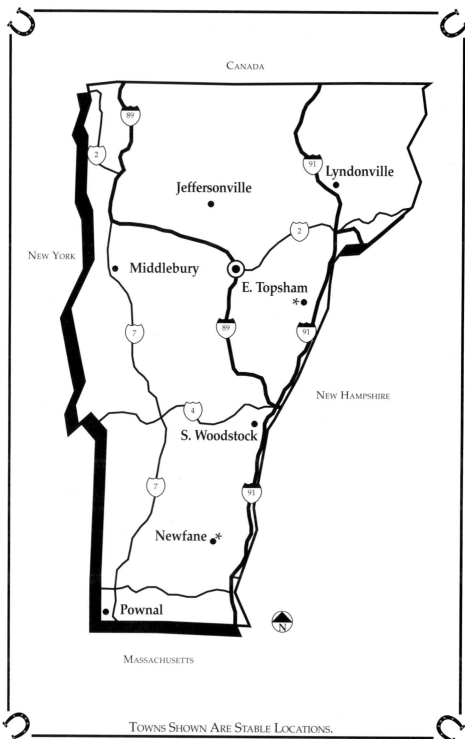

VERMONT PAGE 231

ALL OF OUR STABLES REQUIRE CURRENT NEG. COGGINS, CURRENT HEALTH PAPERS, & OWNERSHIP PAPERS.

EAST TOPSHAM
* Back-in-Time "Horse Resort" Phone: 802-439-5448
Glenn & Burnice Dow
Main Street [05076] **Directions:** I-91 to Exit 16, Rte. 25 west for 7.1 miles, turn right at East Corinth General Store, continue 5 miles to Old Millers Store. **Facilities:** Three 10' x 12' stalls, 1 - 8' x 10' stall, 1 - 24' x 24' stall, two 2-acre pastures, wooden paddock, feed/hay, trailer parking. Historic store/warehouse. Guided trail rides on old logging roads/dirt roads. **Rates:** $20 per night, $100 per week. **Accommodations:** Completely furnished apartment on premises, $75 per day per person/horse. Motels in Bradford, 12 miles away.

JEFFERSONVILLE
Lajoie Stables Phone: 802-644-5347
Amanda Lajoie
992 Pollander Road [05464] **Directions:** Call for directions. **Facilities:** 14 standing stalls, run-in shed, 3-20 acre pastures, 100' x 150' outdoor arena, guided trail rides. Call for reservations. **Rates:** $20 per day. **Accommodations:** Motel located 2 miles from stable.

LYNDONVILLE
Breezy Knoll Stable Phone: 802-626-9685
Harold & Nancy Dresser
RFD 1, Box 361 A [05851] **Directions:** 2.5 miles north of Rt. 5 off of I-91. Call for directions. **Facilities:** 30 indoor box stalls, 70' x 150' indoor arena, 4 paddocks, 2 large fenced pasture areas. Trail riding. Trains Standardbreds for racing. Call for reservations. **Rates:** $15 per night; $75 per week. **Accommodations:** Motel 1 mile from stable.

MIDDLEBURY
Cobble Hill Farm Phone: 802-388-7027
Peggy Ward
RD 3, Painter Road [05753] **Directions:** Call for directions. **Facilities:** 23 indoor stalls, no pasture/turnout, feed/hay & trailer parking available. Call for availability & reservations. **Rates:** $15 per day. **Accommodations:** Many nearby.

NEWFANE
* West River Lodging & Stables Phone: 802-365-7745
Roger Poitras
RR 1, Box 695 [05345] **Directions:** Exit 2 off of I-91 on Rt. 30N. Call for directions. **Facilities:** 24 indoor stalls, 3 paddocks, 160' x 200' outdoor ring. Jumping & dressage lessons up to Grand Prix level based on centered riding. Call for reservations. **Rates:** $20 per night. **Accommodations:** Bed & Breakfast on premises.

VERMONT

ALL OF OUR STABLES REQUIRE CURRENT NEG. COGGINS, CURRENT HEALTH PAPERS, & OWNERSHIP PAPERS.

POWNAL
Valleyview Horses & Tack Shop, Inc.   Phone: 802-823-4649
Shelley Porter
24 Poormans Road [05261] **Directions:** Located off of Rt. 7. Call for further directions. **Facilities:** 7 box stalls, 9 straight stalls, 100 acres fenced pasture, outdoor run-in sheds with 2 horses to each paddock, 80' x 100' outdoor ring, miles of trails on 250 acres. Full tack shop with Western apparel on premises. Buys & sells horses and lessons in English & Western. Trail rides, pony rides, & parties. Call for reservations. **Rates:** $20 per night; $80 per week. **Accommodations:** Many motels 2 miles from stable.

SOUTH WOODSTOCK
Green Mountain Horse Association   Phone: 802-457-1509
Joe Silva, General Manager
Rt. 106 South [05071] **Directions:** Call for directions. **Facilities:** 136 indoor stalls, dressage arena, cross-country course with jumps. Unlimited trail riding. Sponsor of yearly events. Call for reservations. **Rates:** $15 per night for members; $25 per night for non-members. **Accommodations:** Many motels in Woodstock, which is a beautiful & historic town with many sites & activities.

NOTES AND REMINDERS

PAGE 234 **VIRGINIA**

TOWNS SHOWN ARE STABLE LOCATIONS.

* onsite accommodations

VIRGINIA   Page 235

<u>*ALL OF OUR STABLES REQUIRE CURRENT NEG. COGGINS, CURRENT HEALTH PAPERS, & OWNERSHIP PAPERS.*</u>

CANA
Tanbark Acres   Phone: 276-755-5191
Carlton and Dee Everhart   Fax: 276-755-2739   Web: www.tanbarkacres.com
240 Tanbark Trail [24317] **Directions:** 4 miles from I-77, 1 1/10 from US #52. Call for Directions. **Facilities:** 10 12x12 rubber-matted indoor stalls. 4 (small but grassy) paddocks, trailer parking, feed/hay available, Friesian breeding farm, new Morton Barn, negative Coggins, health records, proof of ownership, New River Trail; 157 miles of walking/riding trails **Rates:** $20 per night. **Accommodations:** Several Hotels and Motels within a few miles of the farm.

EASTVILLE
* **Windrush Farm Bed & Breakfast**   Phone: 757-678-7725
Eleanor Gordon   Fax: 757-678-5577
Web: www.windrushholidays.com   E-mail: lcgordan@esva.net
5350 Willow Oak Rd [23347] **Directions:** Right off Rt 13, major N/S Hwy. 20 miles North of Chesapeake Bay Bridge Tunnel. 1 hour North of Norfolk VA and 2 hours South of Salisbury MD. **Facilities:** 5 box stall in big old barn with wash rack, paddocks and pasture. Bring own feed/hay miles of trails including Chesapeake Bay beaches. **Rates:** Call for rates. **Accommodations:** Farmhouse with B&B with 2 rooms and shared bath w/AC. Pets welcome.

FORT VALLEY
* **Fort Valley Stables**   Phone: 540-933-6633
Rick & Sandy Deschenes   Toll Free Phone: 888-754-5771
Web: www.fortvalleystable.com
299 South Fort Valley Road [22652] **Directions:** Exit 279 off I-81. east 1 mile to Hwy 11 - Left. 1/2 mile to Hwy 675 - Right. 5-1/2 miles to Kings Crossing - Right. Hwy 678, 1-1/2 miles to stable entrance. **Facilities:** 8 barn stalls, 22 12'x16'corral pens, 18 water and electric sites, 26 no hookup sites, 3 fishing ponds, over 80 miles of mountain trails, 3 paddocks about 1/4 acre each. **Rates:** $5 to $15 per horse per night. **Accommodations:** 2 cabins on site. Trailer or tent camping with water and electric hookups. Ramada Inn & Budget Host, both in Woodstock 9 miles away.

FREDERICKSBURG
Cedar Crest Farm & Stables   Phone: 540-752-7302
314 Poplar Road [22406] **Directions:** I-95 to rt. 17. Take 4 miles to Poplar Rd. on right(rt. 616). 1 mile to farm sign, take right. Glendie Beside Drive, Follow road to barn. **Facilities:** 8 10x16, 16x16 indoor stables. trailer parking, electric hookups, feed/hay available, 4 separate paddocks, full size riding ring, pasture, year-round full care boarding available. **Rates:** stall-$20 pasture-$10 per night. weekly rates negotiable. **Accommodations:** Holiday Inn, Motel 6, Court Yard, Ramada, EconoLodge within 5 miles of barn.

ALL OF OUR STABLES REQUIRE CURRENT NEG. COGGINS, CURRENT HEALTH PAPERS, & OWNERSHIP PAPERS.

GREENVILLE

* Penmerryl Farm/The Equestrian Centre Phone: 540-337-0622
Ken Pittkin, Owner; Karen Evans, Manager
662 Greenville School Road, Box 402 [24440] Fax: 540-337-0282
Directions: 10 minutes from I-81, Exit 213A Greenville. Call for directions, reservations and rates. **Facilities:** 20-34 indoor 12' x 12' stalls, paddocks, large pastures, horse walker, trailer parking. Farm is working breeding and training center. Also on premises tennis court and two lakes for swimming, fishing, sailing. Must make reservations. **Rates:** $25 per night. **Accommodations:** B&B lodge & cabins with pool and hot tub.

KENTS STORE

Another Bay Farm Phone: 804-457-3408
John Pearsall Blants
4375 Hickory Hill Road [23084] Directions: Take Hadensville Exit off off I-64. Go .5 mile & cross over Rte. 250. Turn left on 606/629 & go 1.5 miles. Turn right on Rt. 609 (Hickory Hill Rd). Go exactly 1 mile then turn left onto farm road. **Facilities:** 7 indoor 12' x 12' box stalls plus 10 outdoor run-ins. 6 pastures on 400-acre farm, riding ring, round pen and miles of trails. Feed/hay and ample trailer parking available. Miles of riding trails. Quarter horses and Paints for sale. **Rates:** $18 per night; $100 per week. **Accommodations:** Motels in Charlottesville area, 20 miles away.

LEXINGTON/NATURAL BRIDGE

Fancy Hill Farm Phone: 540-291-1000
Patricia A. Magner, Manager Fax: 540-291-4057
Web: www.fancyhillfarm.com E-mail: fancyhillfarm@yahoo.com
100 Equus Route [24578] Directions: From I-81 South, take Exit 180B (Route 11, Fancy Hill), turn right on Rte 11, 6/10 mile to entrance on left. From I-81 North, take Exit 180 to Rte 11 North, left on 11, 1.2 miles to entrance on left. **Facilities:** 88 stalls. 34 block 12x12, 54 wood 11.5x11, all indoors. Several 4 5 acre paddocks. All enclosed indoor arena, 4 riding rings, cross country course, 140 acres of trails. Shows, clinics, and sales. Feed and hay available. Easy access for large trucks and trailers. **Rates:** $35 per night, $200 weekly, plus bedding. **Accommodations:** Westmoreland Budget Motel (1/2 mile), Natural Bridge Motel (4 miles). Several motels in Lexington, 7 miles away.

VIRGINIA Page 237

ALL OF OUR STABLES REQUIRE CURRENT NEG. COGGINS, CURRENT HEALTH PAPERS, & OWNERSHIP PAPERS.

LOCUST DALE
* The Inn at Meander Plantation Phone: 540-672-4912
Suzanne Thomas, Suzie Blanchard, & Bob Blanchard
2333 James Madison Hwy [22948] **Directions:** I-81 to I-64 to Rt. 15. 9 miles south. Adjacent to Robinson River. **Facilities:** 5 to 10 indoor stalls in 3 buildings, 60 acres of pasture/turnout, outside riding ring, riding trails, feed/hay at add'l cost, trailer parking. Kennels available for dogs. This is a vacation and sightseeing area. Overnight boarding only for guests of inn. Advance notice preferred. Innkeepers will assist you in finding local trails & areas to ride in the mountains. **Rates:** $20 per night; weekly rate upon request. **Accommodations:** Inn on premises in a stately Colonial mansion. Five guest rooms with private baths and full breakfast. Only a short drive to Skyline Drive, Charlottesville & Monticello, & Montpelier.

MIDDLEBURG
Middleburg Equine Swim Center Phone: 540-687-6816
Roger Collins & Laura Hayward
35469 Millville Road [22117] **Directions:** 3 miles west of Middleburg on Rte. 50, turn right on Rte. 611, after 1 mile turn right onto Millville Road. Swim Center is 1 mile on right. **Facilities:** 42 indoor 12' x 12' stalls, pasture/turnout of varying sizes, 100' x 200' outdoor riding ring. feed/hay, trailer parking. Unique facility specializing in swimming horses for rehab and conditioning. In the heart of Virginia, foxhunting on 42 acres, miles of trails and cross-country riding. **Rates:** $15 per night. **Accommodations:** Numerous B&Bs and country inns within 3 miles.

MIDDLETOWN
Monte Vista Stable Phone: 540-869-4621
Dr. N. Lee Newman Fax: 540-869-0979
8183 Valley Pike [22645] **Directions:** Call for directions. Located 2 miles from I-81 and I-66 intersection. **Facilities:** 6 indoor stalls minimum 10' x 14', small paddocks, feed/hay, trailer parking. National historic register property; veterinarian on premises; Rowdi Arabians, foals for sale; near National Forest trails, Old Dominion 100 trails. Reservations required. **Rates:** $20 per night, $100 per week. **Accommodations:** Bed & Breakfast available on site. Call for prices. Wayside Inn in Middletown, 1 mile; Comfort Inn in Stephens City, 5 miles; Battle of Cedar Creek Campground in Middletown, 1 mile. Super 8 Motel 1.5 miles.

PETERSBURG
Idle Moment Farm Phone: 804-862-4463
Garry G. & Bobbie L. Moretz
7724 Vaughan Road [23805] **Directions:** Call for directions. **Facilities:** 15 indoor stalls, 66' x 166' arena, 40 x 40' paddock, dressage arena, feed/hay & trailer parking. Recent negative coggins and health certificate "We'll be happy to accommodate almost any request." **Rates:** $20 per night. **Accommodations:** Several motels & restaurants within 5 miles.

ALL OF OUR STABLES REQUIRE CURRENT NEG. COGGINS, CURRENT HEALTH PAPERS, & OWNERSHIP PAPERS.

POWHATAN
Allengeny Stables   Phone: 804-379-2970
Ron Ervin
1735 Old Powhatan [23139] **Directions:** Located 22 miles from the center of Richmond on Rt. 60W. Call for directions. **Facilities:** 4 indoor stalls, 4 acres of pasture/turnout, feed/hay & trailer parking available. Limited trail riding and pony rides available. Call for reservation. **Rates:** $20 per night; ask for weekly rate. **Accommodations:** Days Inn 7 miles from stable.

Windsor Farm Stables   Phone: 804-598-2679
Joe Hairfield
2600 Huguenot Trails [23139] **Directions:** Located off of Rt. 60. Call for directions. **Facilities:** 4 indoor stalls, 240 acres of pasture/turnout, feed/hay & trailer parking available. 2 days notice if possible. Major tourist attractions nearby. **Rates:** $15 per night includes feed. **Accommodations:** Motels within 15 miles of stable.

QUINTON
* **The Winged Horse**   Phone: 804-932-9285
David M. Ruslander
2949 Pocahontas Trail [23141] **Directions:** 1/2 way between Richmond and Williamsburg, VA. Off I-64: exit 205, south to Bottoms bridge, Rt. 60/ Pocahontas Trail Take left onto Rt. 60, 1.1 miles on right. **Facilities:** 3 indoor 12'x12' stalls, Nelson automatic waterers, ceiling fans, rubber mats, auto-feeders. Feed/hay available, trailer parking, various permanent and temporary pastures with Nelson auto waterers. Barn has infrared heaters, hot & cold water, wash stalls, and washer and dryers for humans. No electrical fencing (all safe board or flexible vinyl over wire fencing). Remote controlled entry gate and full-time farme manager on premises. Small animal hospital nearby, non-smokers preferred. **Rates:** Call for Rates. **Accommodations:** On Site: rooms with private baths, studio apartment-wheelchair accessible. Separate entrances, Satelliet TV, separate HVAC, outdoor heated inground pool.

RADFORD
Bedlam Manor Farm   Phone: 540-639-4150
Rebecca Thompson   E-mail: fraubedlam@aol.com
6363 Belspring Rd (Mail: P.O. Box 3425) [24143] **Directions:** From I-81 S: Take Exit 109; bear right onto Rt. 177/Tyler Ave. & continue to Norwood St.; turn left on Rt. 11 (Norwood St.) & follow Rt. 11 S, turning right to cross Memorial Bridge; turn right onto Rt. 114 at 2nd stop light; turn left onto Rt. 600 at first stop light. Farm is one mile on left. **Facilities:** 4 large box stalls, turnout paddocks available with shelter, feed/hay available with prior notice, trailer parking on farm. Located reasonable distance from trails in Jefferson National Forest and the New River Trail. Neg. Coggins within 60 days, health records up to date. **Rates:** $20 per night for one horse; $10 per night each for more than one. **Accommodations:** Dogwood Motel & Executive Motel both 5 miles from stable.

VIRGINIA   Page 239

ALL OF OUR STABLES REQUIRE CURRENT NEG. COGGINS, CURRENT HEALTH PAPERS, & OWNERSHIP PAPERS.

SALEM
<u>Sundance Manor</u>　　　　　　　　　　　　　Phone: 540-380-4001
LaClaire Dantzler, trainer
5091 Glenvar Heights Blvd. [24153]   **Directions:** Dixie Caverns Exit off of I-81. Call for directions. **Facilities:** 11-15 indoor stalls, outdoor paddocks, pasture/turnout area, feed/hay & trailer parking at stable. Teaching, training, & breeding done at stable. Specializes in American Saddlebred. As much advance notice as possible. **Rates:** $30 for stall, $15 for paddock per night. **Accommodations:** Blue Jay & Super 8 less than 10 minutes from stable.

STANLEY
* <u>Jordan Hollow Farm Inn</u>　　　　　　　　Phone: 540-778-2285
Millie Short　　　Web: www.jordanhollow.com　　　540-778-2209
326 Hawksbill Park Road (SR 626) [22851]   **Directions:** From I-81 at New Market: Take Rt. 211 east to Luray; turn on 340S <u>Business</u> & go 6.5 miles to left on Rt. 624 to stop sign; take left on Rt. 689 & go approx 1/2 mile; take right on Rt. 626. Farm is .3 mile on right. **Facilities:** 10 indoor stalls, 2 small paddocks, riding ring, small field, timothy/grass mix at $2.50 per bale. **Rates:** $20 for non-guests of inn; $15 for guests. **Accommodations:** Jordan Hollow Farm Inn is a beautifully restored colonial horse farm that has been converted to a country inn. It has 20 guest rooms and a restaurant on the property. $110 or $154 per night for 2 people including breakfast.

STAUNTON
<u>Cedar Creek Stables</u>　　　　　　　　　　Phone: 540-294-3003
Melyni Worth
188 Sherwood Dr. [22980]   **Directions:** 3 miles off of I-81 & 2 miles off I-64. Call for directions. **Facilities:** 40 indoor stalls, 29-acre pasture, 1.5 & 2.5-acre paddocks, 75' x 50' outdoor ring, inside riding available. English & Western lessons offered. Horses for sale. Call for reservation. **Rates:** $20 per night. **Accommodations:** Holiday Inn, Days Inn & Master Host Motel 2 miles from stable.

<u>Westwood Animal Hospital</u>　　　　　　　Phone: 540-337-6200
Susan Trout, manager
Rt. 6, Box 453A [24401]   **Directions:** 5 miles from I-64 & I-81. Call for directions. **Facilities:** 6 indoor stalls, 3.5 acre pasture, turnout facilities, outdoor riding ring, feed/hay & trailer parking on premises. Large animal care available. Call for reservation. **Rates:** $20 per night. **Accommodations:** Holiday Inn & Master Host Hotel 5 miles from hospital.

VIRGINIA

ALL OF OUR STABLES REQUIRE CURRENT NEG. COGGINS, CURRENT HEALTH PAPERS, & OWNERSHIP PAPERS.

WARRENTON
✳ Black Horse Inn   Phone: 540-349-4020
Lynn Pirozzoli   Web: www.blackhorseinn.com
8393 Meetze Road [20187] **Directions:** Call for Directions **Facilities:** 9 10'x12' stalls, trailer parking, pasture. **Accommodations:** Bed & Breakfast on premises. See website www.blackhorseinn.com, Comfort Inn and Hampton Inn within 10 miles.

WASHINGTON
✳ Caledonia Farm - 1812   Phone: 540-675-3693
Phil Irwin   Reservations: 800-BNB-1812
47 Dearing Road [22627] **Directions:** Call for directions. 4 miles north of Washington, Virginia, and 68 miles southwest of Washington, D.C. Close to I-66, I-81, I-95, and I-64. **Facilities:** 2 indoor barn stalls, 50' x 50' pasture/turnout on 52 acres, trailer parking. Adjacent to Shenandoah National Park with its 500 miles of trails. Western & English stables/studs nearby. **Rates:** Free to B&B guests. **Accommodations:** B&B on premises.

WILLIAMSBURG
Carlton Farms   Phone: 757-220-3553
C. Lewis Waltrip
3516 Mott Lane [23185] **Directions:** I-95 to I-64 E. Call for further directions. **Facilities:** 10 indoor stalls, huge indoor ring, pasture/turnout area, feed/hay & trailer parking available. A boarding & lessons facility that also sponsors horse shows. 1 day notice if possible. **Rates:** $20, including feed, per night. **Accommodations:** Motels in Colonial Williamsburg 10 minutes.

NOTES AND REMINDERS

Page 242 WASHINGTON

Towns Shown Are Stable Locations.

* onsite accommodations

WASHINGTON Page 243

ALL OF OUR STABLES REQUIRE CURRENT NEG. COGGINS, CURRENT HEALTH PAPERS, & OWNERSHIP PAPERS.

ARLINGTON
Bill & Stevie Somes Phone: 360-435-3374
6007 267th Place N.E. [98223] **Directions:** From I-5 north of Everett, WA: Take Exit 212; go east on Stanwood-Bryant Rd.; go 4.5 miles to stop sign on Hwy 9; go thru stop sign onto Grand View for 1/2 mile; go right on 59th St.; stable is first driveway on left - red house & white barn. **Facilities:** 2 indoor stalls, 4 outdoor stalls. All stalls have 300' x 40' paddocks. Feed/hay & trailer parking available. 14,000 acres of riding trails adjacent to stable & 64,000 acres available within 3 miles. Access to trailhead for Pacific Crest Trail 90 minutes away. 30 minutes from Puget Sound. **Rates:** $12 per night; $52.50 per week. **Accommodations:** Arlington Motel on I-5 & Hwy 530 about 3.5 miles away.

CHENEY
Marshland Equestrian Center Phone: 509-448-0681
Carolynn Bohlman, Owner or: 509-448-0466
12711 S. Gardner Rd. [99004] **Directions:** Call for directions & availability. **Facilities:** 40 stalls, covered outdoor pens & paddocks. 1 official outdoor Dressage Arena, 1 outdoor Jumping Ring & 1 indoor arena with heated lounge. Quality hay & grain. Trailer parking on premises. Boarding, training, lessons, schooling horses & sales **Rates:** $20 per night. **Accommodations:** Hampton Inn & Quality Inn 10 minutes from stable.

DEER PARK
Blue Haven Stables Phone: 509-276-7968
Randy & Pamela Heiman
W. 8516 Staley Rd. [99006] **Directions:** Located 4 miles west of Hwy 395 at Staley Road Exit. Call for directions. **Facilities:** 19 indoor stalls, indoor arena, 40 acres of pastures & paddocks, 1/4-mile outdoor track, feed/hay & trailer parking available. Has jogging machine for horses that goes up to 18 mph for race horses. Breeds, trains, & sells American Saddlebreds. Farrier on premises. Call for reservation. **Rates:** $15 per night; $75 per week. **Accommodations:** Motels 6 miles from stable.

LONG BEACH
Red Barn Arena Phone: 354-642-2541
Amy Mcttale E-mail: pss@willapabay.org Fax: 360-642-4757
6409 Sandridge Rd. [98631] **Direction:** Call for directions. **Facilities:** 27 indoor stalls. Feed/hay and trailer parking available. We are located 1 mile from 32 miles of beach riding. **Rates:** $12 per night. **Accommodations:** Call for information.

OLYMPIA
James Gang Ranch Phone: 360-491-3216
Linda James
8935 Mullen Rd. SE [98513] **Directions:** Take Exit 111 off of I-5. Call for further directions. **Facilities:** 30-35 indoor stalls, turnouts, feed/hay included, trailer parking on premises. Boarding, training, & breeding facility specializing in Pintos, American Saddlebreds, & Arabians. 1 day notice if possible. **Rates:** $10 per night. **Accommodations:** Motel 8, Olympic Motel, & Quality Inn 3 miles from stable.

PAGE 244 **WASHINGTON**

ALL OF OUR STABLES REQUIRE CURRENT NEG. COGGINS, CURRENT HEALTH PAPERS, & OWNERSHIP PAPERS.

OLYMPIA
Weeping Willow Ranch   Phone: 360-491-3217
4437 Shincke Road NE [98506] **Directions:** Located off of I-5. Call for directions. **Facilities:** 2 indoor stalls, paddocks, indoor & outdoor arenas, feed/hay, trailer parking on site, riding trails. Call for reservation. **Rates:** $20 per night; weekly rate available. **Accommodations:** Many motels in Olympia, 4 to 5 miles from stable.

PORT TOWNSEND
Jefferson County Fairgrounds   Phone: 360-385-1013
Bob Bates
49th & Kuhn [98368] **Directions:** Take Hwy 20 into Port Townsend. Look for signs to fairgrounds, 1 mile but in city limits. **Facilities:** 75 covered all-wood 12' x 12' stalls, 50' x 150' outside riding arena, small inside 70' x 70' arena, feed/hay, trailer parking. Caretaker 24 hours a day. Camping, travel trailer park, dance hall, dining room & kitchen available for rent. Must clean own stalls. **Rates:** $5 per night, weekly rate available.

POULSBO
Sandamar Farm   Phone: 360-779-9861
Reg & Julie Gelderman
4499 NE Ganderson Road [98370] **Directions:** Call for directions. 45 minutes north of Takoma, 1 hour west of Seattle. **Facilities:** 10 indoor 12' x 12' stalls, pasture/turnout, feed/hay, trailer and RV camper parking available. Close to Olympic National Park, guided trail rides available. Call for reservations. **Rates:** $15 per night. **Accommodations:** Motels within 3 miles.

SPOKANE
Spokane Sport Horse Farm   Phone: 509-448-3722
Christel Carlson   Advance Res/Leave message - Barn: 509-448-5064
Message/Fax: 509-448-2658   Cell/ Day of Stay Res: 509-993-6786
E-mail: sales@spokanesporthorse.com   Web: www.spokanesporthorse.com
10710 S. Sherman Road [99224] **Directions:** From I-90, take Pullman Hwy (195) south approximately 3 miles, right on Cheney-Spokane Road, keep right at Y in road, 1 mile past cemetery take left on Sherman Road. Farm is 1.8 miles on right. **Facilities:** 60 indoor stalls with runs, 3 covered stalls with runs, 70' round pen for turnout, 150' x 250' and 100' x 230' outdoor arenas, 216' x 122' indoor arena, automatic heated waterers, feed/hay, trailer parking. Miles of trails on 150 acres. Sale horses; Warmbloods and Lipizzan crosses. Five annual dressage shows. Lipizzan Stallion Conversano ll Natasha at stud. "Deep Creek" feed and tack store on farm premises. **Rates:** $20 per night. $350 monthly. **Accommodations:** Electric hook-up for campers. Hampton Inn, Ramada Inn, and others within 10 minutes.

WASHINGTON   Page 245

ALL OF OUR STABLES REQUIRE CURRENT NEG. COGGINS, CURRENT HEALTH PAPERS, & OWNERSHIP PAPERS.

STANWOOD
Foggy Hollow Farm   Phone: 360-629-3937
Sharon & Al Gileck
5031 - 324th Street N.W. [98292]   **Directions:** From I-5 North: Take Exit 215, 300th St. NW; from ramp, take left; go under freeway to stop; go right on Old 99; follow that for about 2 miles; at 324th St. NW, take left & farm is 2nd house on right. Call for directions from I-5 South. **Facilities:** 6 indoor 12' x 12' stalls, 60' round pen, 50' x 150' & 150' x 200' pasture/turnout areas; grain & hay available, trailer parking on premises & electric & water hook-up for RV. Vet on call 24 hours. **Rates:** $15 per night; $85 per week. **Accommodations:** Hill Side Motel in Conway, about 4 miles from stable.

YAKIMA
White Birch Stables   Phone: 509-452-3184
Roger & Sue Hart
Ray Symmonds Road [98901]   **Directions:** Exit 26 off of I-82. Call for easy directions. **Facilities:** 40 indoor stalls, 50 outdoor stalls, pasture & turnout, indoor arena, 2 lighted outdoor arenas, feed/hay at extra charge, trailer parking available. RV hook-ups and showers available. Natural Grow Grain Distributor. Advance notice if possible but can take on short notice. **Rates:** $10 per night; $15 per night including feed. **Accommodations:** Days Inn nearby. Call stable in advance for discount rate.

Page 246 WEST VIRGINIA

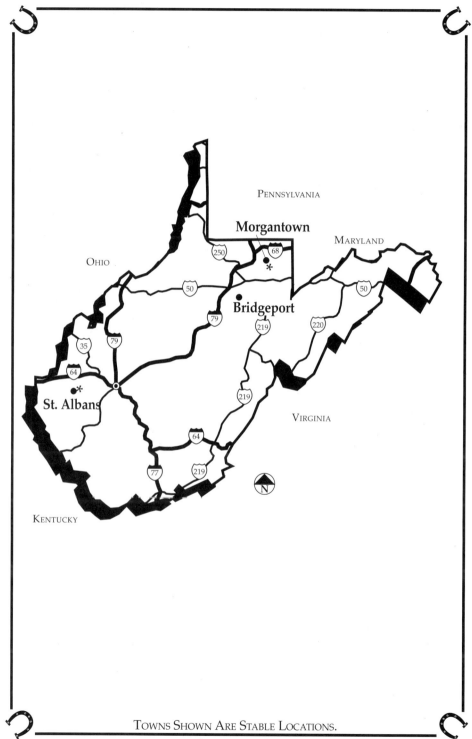

Towns Shown Are Stable Locations.
* onsite accommodations

WEST VIRGINIA   Page 247

ALL OF OUR STABLES REQUIRE CURRENT NEG. COGGINS, CURRENT HEALTH PAPERS, & OWNERSHIP PAPERS.

BRIDGEPORT
4-T Arena Phone: 304-592-0703
Jeff & Larry Tucker   Phone: 304-592-3161   Phone: 304-641-1681
Rt. 3, Box 242F [26330]   **Directions:** 4 miles off of I-79. Call for directions. **Facilities:** 10 indoor stalls, eight 20' x 40' turnout paddocks, 250' x 85' indoor arena, 300' x 200' outdoor arena, feed/hay & trailer parking on premises. Rodeos, shows, & clinics held at arena. Camper hook-up. Vet and farrier on premises. Facilities available to rent. Easy access & large parking lot. Call for reservation. **Rates:** $15 per night. **Accommodations:** Red Roof Inn-Fairmont (6 miles), Hampton Inn-Bridgeport (7 miles).

MORGANTOWN
✶ **Appelwood Bed & Breakfast and Stables**   Phone: 304-296-2607
Jim Humbertson   Web: www.appelwood.com
1749 Smithtown Road, RR 5 Box 137 [26508]   E-mail: appelwood@aol.com
Directions: Exit 146 off of I-79 to Goshen Road, 100 yds to stop sign at Junction 73. Turn right and go 1.5 miles. Driveway is on right across from white mailbox. **Facilities:** 9 indoor 12' x 11' stalls; 1-, 3-, and 5-acre pastures; ring; feed/hay available; trailer parking. Some trails on the 35-acre premises. **Rates:** $15 per night. **Accommodations:** Bed & Breakfast on premises, $85 per night.

Meadow Green Stables   Phone: 304-296-1979
Ken & P.J. Neer & Tim
Rte. 10, Box 161D [26505]   **Directions:** Exit 68E off I-79. 5-10 minutes off I-68. University Avenue Exit off of I-68. Left off interstate onto 119. Go 2 lights, right on 857. Straight at next light. Right at 4-way stop onto Kingwood Pike. 1.1 mile on left. 5 minutes from Mountaineer Mall; 10 minutes from WV University. **Facilities:** 24 - 12' x 12' stalls, 5' high board fenced paddocks, indoor/outdoor arena, individual fans, individual diet & exercise needs catered to, feed/hay, trailer parking, climate controlled lounge, boarding & training, lessons in beginner & advanced English. Warm water baths. No stallions. Coggins & shot records will be checked & are required. Please do not untrailer before papers are checked. Call for reservations. **Rates:** $25 per night. **Accommodations:** Ramada Inn, Comfort Inn within 3 miles.

ST. ALBANS/ CHARLESTON AREA
✶ **Sunday Stables**
Susan & Christy Sunday, Owner   Phone: 304-722-4630
E-mail: ssunday@access.k12.wv.us
1 Twilight Lane [25177]   **Directions:** St. Albans exit of I-64 (west of Charleston). Right at bottom of ramp to rt 35. Turn left onto rt 60 at stop light. Go 1 mile to Subway Sandwich shop and go 3 miles. Bear left through tunnel and take immediate right. Look for Sunday Stables sign. **Facilities:** 25 10x12 inside stalls, inside arena. trailer parking, feed/hay available, training, lessons, sales. **Rates:** $20 per night. **Accomodations:** Bed and Breakfast starting at $50 per night.

PAGE 248 WISCONSIN

Towns Shown Are Stable Locations.

* onsite accommodations

WISCONSIN Page 249

ALL OF OUR STABLES REQUIRE CURRENT NEG. COGGINS, CURRENT HEALTH PAPERS, & OWNERSHIP PAPERS.

BARNEVELD
Black Forest Farm, LLC
Marlene Cordes
Phone: 608-437-4905
Web: www.blackforeststables.com
E-mail: only@blackforeststables.com
3255 County Road F [53507]
Directions: 3 miles South of Hwy 18/151 on County road F at Blue Mounds. 30 minutes West of Madison. **Facilities:** 21 stalls available (12x12, 12x24, 12x20, 10x24) some with individual runs. Feed/hay available, trailerparking with electric and water hookup, several 1-10 acre pastures, Standing at Stud, quarter horse black son of Impressive, IM Magic. Breed and sell Paints and Quarter Horses. Please Call ahead for reservations. **Rates:** Stall board varies with size and amenities of stall. Call for rates **Accommodations:** Numerous motels within 30 minutes.

Black Forest Farm LLC
Premium Horses of Color and Performance
Quarter Horses   Paint Horses

Home of black Quarter Horse stallion **IM Magic**
Quality Horses and Foals Available for Purchase
• Breeding   • Foaling Suites
• Boarding   • Horse Motel
• Training for Horses & People
• Magnum & Equilitter Bedding Dealer

Only 30 minutes from Madison's West Side 3255 County Road F, 3 Miles South of 18/151 at Blue Mounds

  608-437-4905

www.blackforeststables.com

CHIPPEWA FALLS
Amber Farm, Inc.   Phone: 715-723-7050
Spencer & Kathy Jerome   715-723-9513
18798 70th Avenue [54729] **Directions:** Call for directions. **Facilities:** 12 indoor box stalls, 10 indoor pipe stalls, feed/hay available, trailer parking, 50'x90' turnout **Rates:** 25 per night **Accommodations:** Americinn -4 miles, Pine Harbor Campground - 1 mile.

COTTAGE GROVE
Lazy L Ranch & Horse Company   Phone: 608-873-6725
Skip & Lila Lemanski
2189 Rinden Road [53527] **Directions:** From I-90, take Exit 147. Call for directions. **Facilities:** 30 indoor stalls, round pen, paddocks, 60' x 120' indoor arena, 2 large outdoor lighted arenas. Monthly boarding & lessons. Owners on premises. Vet on call. Convenient to Madison & Wisconsin University Veterinary School. Call for reservation. **Rates:** $20 per night; call for weekly rate & availability. **Accommodations:** Motel 6 miles from stable.

ALL OF OUR STABLES REQUIRE CURRENT NEG. COGGINS, CURRENT HEALTH PAPERS, & OWNERSHIP PAPERS.

HOLCOMBE

* **Merrimount Stables & Happy Horse B&B** **Phone: 715-239-6158**
Alan & Sandra Ricker
24469 State Hwy 27 [54745] Directions: On Hwy 27 approximately 50 miles North of Osseo/I-94 exit, or 150 miles East of Twin Cities. **Facilities:** 4 10'x12' indoor stalls, 100'x100' pasture, feed/hay available, trailer parking, registered Morgans-Standing Alson Paragon. **Rates:** $10 per night. **Accommodations:** Bed & Breakfast owned by Alan & Sandra Ricker on premises.

KNAPP

Pinehaven **Phone: 715-643-6018**
William & Margaret Miller
E1901 890th Ave [54749] Directions: Call for directions. **Facilities:** 6 indoor box stalls. Feed available. Pasture, turnout available. Trailer parking. Trail riding available. **Rates:** $20 per night. $100 per week. **Accommodations:** American Inn, Best Western, Super 8, Motel 6 all in Menomonie 10 miles away.

RACINE

Sundown Stables **Phone: 262-878-9270**
Joe Stang
4218 7 Mile Road [53402] Directions: Located conveniently off of I-94. Call for directions. **Facilities:** 40 indoor stalls, pasture/turnout area, feed/hay included, trailer parking. Open 7 days a week: 8 A.M. to 7 P.M. Horseback riding and hay rides offered at stable. Call for reservation. **Rates:** $25 per night; weekly rate negotiable. **Accommodations:** Motels within 5 miles of stable.

SHEBOYGAN

Pine Rock Stables **Phone: 262-458-1113**
Pam Becker
1431 County Hwy V [53081] Directions: From I-43: Exit east on Hwy V; follow hwy. about 3/4 mile to near the end; on the right, go down driveway to barn in back. **Facilities:** 23 indoor stalls, 10 outdoor stalls with run-in shelters, 25' x 10' turnout paddocks, 3 acres of fenced pasture, 50' x 150' indoor arena, 130' x 150' outdoor arena, round pen & jump course. Located next to
Kohler-Andrae State Park with campsites & riding trails. This is a boarding & training facility offering beginning lessons in English & Western and advanced lessons in Western pleasure, dressage, hunt seat, & jumping. Horses that crib must be restrained in some manner. Call for reservation. **Rates:** $15 per night; $75 per week. **Accommodations:** Camping in state park; Parkway Motel in Sheboygan, 1 mile from stable.

WISCONSIN Page 251

ALL OF OUR STABLES REQUIRE CURRENT NEG. COGGINS, CURRENT HEALTH PAPERS, & OWNERSHIP PAPERS.

SPRING GREEN
Endless Valley Stables Phone: 608-753-2887
Steven Murphy Web: www.endlessvalleystables.com
E-mail: info@endlessvalleystables.com
5975 Hwy T [53588] **Directions:** Call for directions. **Facilities:** 10 10'x12' box stalls. Feed/hay available, trailer parking available. Private trails, training, lessons and day camps. **Rates:** $15 per night, monthly board available. **Accommodations:** Camping and lodging with full kitchen, public restrooms with shower. Many motels in Spring Green and Dodgeville.

WHITEWATER
Echo Valley Farm Phone: 262-473-4631
Thomas P. & Rhonda L. Fuller Fax: 262-473-4635 E-mail: fuller@idcnet.com
W3218 Piper Road [53190] **Directions:** Off state Hwy 59 between Madison and Milwaukee. Call for specific directions. **Facilities:** 3 indoor 10' x 10' box stalls, 18' x 10' enclosed stall with attached 100' x 125' paddock, two 2-acre pastures, all access to 18' x 28' enclosed building for shelter. All fencing "Country Estate 3-Rail PVC." Feed/hay and trailer parking available. Located between Milwaukee, Madison & Janesville. Six miles from Horseman's Park, located in Southern Kettle Moraine State Forest with 50 miles of riding trails available. Standing stud Clydesdale stallion, Bluffviews' Sir Ethan (Ayton Perfection Breeding). **Rates:** $15-$25 per night. **Accommodations:** Super 8 Motel and Amerihost Inn in Whitewater, within 6 miles. Best Western in Fort Atkinson next to the nationally known Fireside Restaurant & Playhouse, within 10 miles. Also, several B&Bs in area.

WYOMING

Towns Shown Are Stable Locations.

* onsite accommodations

WYOMING PAGE 253

ALL OF OUR STABLES REQUIRE CURRENT NEG. COGGINS, CURRENT HEALTH PAPERS, & OWNERSHIP PAPERS.

BIG HORN

✳ **Bozeman Trail Bed & Breakfast**   Phone: 877-672-2381
Marie Johnson
304 Hwy 335 (Mail: P.O.Box 608) [82833] **Directions:** From I-90 take Exit 25, go west to Coffeen Ave., turn south/left on Hwy 87, 4 miles to Texaco at the Y, bear right onto Hwy 335, stable 3 miles. **Facilities:** 7 indoor stalls, 4 with outdoor runs, 1 foaling stall, two 1-acre turnouts, feed/hay, trailer parking. Close to Big Horn Equestrian Center & Big Horn National Forest; King Saddlery 10 miles . Reservations required. **Rates:** $15 per night. **Accommodations:** B&B on premises. Days Inn, Holiday Inn in Sheridan, 9 miles away.

CASPER

Central Wyoming Fairgounds   Phone: 307-235-5775
Tom Jones, Manager   Web: www.centralwyomingfair.com/
E-mail: cwfr@centralwyomingfair.com
1700 Fairgounds Road [82604] **Directions:** Call for Directions **Facilities:** 511 10'x10' outdoor stalls **Rates:** $5 per night. **Accommodations:** Many Lodging Options nearby.

CHEYENNE

✳ **A. Drummond's Ranch B & B**   Phone/Fax: 307-634-6042
Kent &Taydie Drummond   E-mail: adrummond@juno.com
Web: cruising-america.com/drummond.html
399 Happy Jack Road (State Hwy 210) [82007] **Directions:** Call ahead for directions. **Facilities:** 4 indoor 12' x 12' stalls with rubber mats, 1 run-in stall, 60' x 120' indoor arena, four 24' x 32' corrals, $5 bale for hay, trailer parking on premises. Will feed your horses in the AM. 55,000 acres for riding in Medicine Bow National Forest nearby. Advanced notice reservations required. Not available for same day accommodations. No smoking. **Rates:** $10- outside corral, $20-stall with bedding, $25- stall, bedding & run. No dogs in rooms, or off leash! **Accommodations:** Bed & Breakfast on premises for horse & hauler. 4 bedrooms sleep up to ten. Quiet, gracious setting of view, wildlife & garden on 120 acres. Reasonable rates vary depending on time of year.

✳ **Adventurers' Country Bed and Breakfast,**   Phone: 307-632-4087
Horse Motel and Raven Run Kennels   E-mail: fwht2@aol.com
Fern White   Web: www.bbonline.com/wy/adventurers/
3803 I-80 S. Service Road [82009} **Directions:** Off Interstate 80 east, take exit #377, take a left, go to Frontage Rd. Take a left, go 1-1/2 miles and see sign that reads "South Service Rd, No Outlet" look for mailboxes and underpass on your left. Go under, bear left take second road to right. Follow around to the long white house, red barn on hill. Park in the front of barn and come to east end of house. **Facilities:** 4 indoor, 12x12 paneled stalls, 2 large paddocks, lots of parking for all types of trailers & vehicles, reservations required. Barn facilities for dogs, we breed and show yellow labs. **Rates:** $15 per night for stall, $7.00 hay, $5.00 bedding. **Accommodations:** Bed and Breakfast(non-smoking, no dogs in room) on site with queen beds, private bathroom and full breakfast.

WYOMING

ALL OF OUR STABLES REQUIRE CURRENT NEG. COGGINS, CURRENT HEALTH PAPERS, & OWNERSHIP PAPERS.

CHEYENNE
<u>Cheyenne Stockyard-Cattle & Horse Motel</u>　　　Phone: 307-634-5696
Duane Iliff (Pappy)　　　Fax: 307-637-4826　　　Toll free: 888-634-5696
Web: www.busdir.com/cheyennest　　　E-mail: cheysock@aol.com
350 Southwest Drive [82009] **Directions:** Call for directions. Within 1 mile of I-80 and I-25. **Facilities:** 21 - 20 X 25, 16 - 20 X 40 indoor stalls, 7 - 40 X 70 outdoor pens, feed/hay, trailer parking, vet/farrier/groomer on call, extended stay available. **Rates:** $15 per night. **Accommodations:** Within walking distance of restaurants, hotels, gas station, truck stop.

✶ <u>7XL Stables at Terry Bison Ranch</u>　　　Phone: 307-634-4171
Web: www.terrybisonranch.com
51 I-25 Service Road East [82007] **Directions:** On I-25 at Wyoming-Colorado border. Take WY Exit 2 to Terry Ranch Rd. South 2 miles on Terry Ranch Road. **Facilities:** 16 indoor stalls, 12 outdoor stalls, outdoor pens available for semi loads, feed/hay available, trailer parking. Overnight and monthly stabling for horses on a 27,000-acre bison ranch. Unique guest ranch experience with horse-drawn wagon tours, fishing, RV park, guest cabins, bunkhouse, restaurant, saloon and gift shop. Year-round accommodations. **Rates:** Varies. **Accommodations:** RV Park, guest cabins, & 17-room bunkhouse on site.

<u>Singletree Stable</u>　　　Phone: 800-336-0287
Doug Terlizzi
4717 Thomas Road [82009] **Directions:** Exit 364 off of I-80. Call for directions. **Facilities:** 14 indoor 12' x 12' box stalls with auto waterers, lighted indoor & outdoor arenas, feed/hay & trailer parking available. **Rates:** $15 per night; weekly rate negotiable. **Accommodations:** Motels 3 miles from stable.

DOUGLAS
✶ <u>Deer Forks Ranch</u>　　　Phone: 307-358-2033
Ben and Pauline Middleton　　　Web: www.deerforks.com
E-mail: deerfork@netcommander.com
1200 Poison Lake Road [82633] **Directions:** Call for directions. 25 miles south of I-25. **Facilities:** Stalls and ranch corrals available as needed, small pastures on 15,000 acres, summer pasture only, winter supplement as needed, trailer parking, RV hook-up. Ranch raises cattle, show sheep & paint horses. Big game hunting in fall. Trail rides and cattle drives available. Advance reservations recommended. **Rates:** $10 per night, $60 per week, $50 per month for pasture. **Accommodations:** B&B on premises.

WYOMING Page 255

ALL OF OUR STABLES REQUIRE CURRENT NEG. COGGINS, CURRENT HEALTH PAPERS, & OWNERSHIP PAPERS.

FORT LARAMIE
Carnahan Ranch
Hal & Bess Carnahan
935 Gray Rocks Rd [82212]
E-mail: carnahanranch@scottsbluff.net

Phone/Fax: 307-837-2917
Toll Free: 800-837-6730
Web: www. carnahanranch.com

Directions: Quiet, secluded private ranch with easy access to all Hwys. At Fort Laramie follow signs to Fort Laramie National Historic Site. Go 1/2 mile past entrance to fort, across bridge, turn left into ranch drive. **Facilities:** 2400 acres bordering Fort Laramie NationalHistoric Site. Stalls, corrals, water, full size arena, alfalfa/grass hay available. Lots of room for trailers and RVs. RVhookup electric and water only. Rec building with 2 full bathrooms, full kitchen, serving counter, games, picnic tables and grills. Vet service nearby. Trail riding on your own on historic ranch. All of the history of the west passed through the ranch, Oregon, Mormon, California, Cheyenne/Deadwood Trails, Indian campgrounds, etc. Groups welcome for cattle drives and or historic rides, camp outs, scout camps, family reunions, weddings, AERC endurance rides. Home to rare breed of gaited Spanish Colonial Mustangs, Longhorn and Angus cattle. **Rates:** Per night- campsites, tent & trailers $15, RVw/electric $20, horses $5, with feed $8. Full use of facilities included. Reservations recommended. Long term rates available on request. **Accommodations:** motels in Torrington.

GILLETTE
* P Cross Bar Ranch
Marion & Mary Scott

Phone: 307-682-3994
Email: pcrossbar@vcn.com

8586 N Hwy 14-16 [82716]   **Directions:** Call for directions. Right on US 14-16, 20 miles north of Gillette. **Facilities:** 3 indoor 15' x 20' stalls, 4 corrals with shed set-up, feed/hay, trailer parking, small fee for camper parking. Big game outfitters. Call for reservations. **Rates:** $10-$15 depending on service; weekly rates available. **Accommodations:** B&B on premises.

GREYBULL
* McFadden Ranch
Sandy McFadden
Web: www.mcfaddenranch.com

Phone: 307-765-9684
Fax: 307-765-9609
E-Mail: training@mcfaddenranch.com

2480 Lane 30 1/2 (82426) **Directions:** Call for directions. **Facilities:** 13 12x12 and 12x24 stalls. Feed/hay and trailer parking available. Pasture and turnout available. **Rates:** $25 per night, $100 per week. **Accommodations:** Apartment available on ranch $45 per night. Motels in Greybull 5 miles from ranch.

WYOMING

ALL OF OUR STABLES REQUIRE CURRENT NEG. COGGINS, CURRENT HEALTH PAPERS, & OWNERSHIP PAPERS.

HOBACK JUNCTION
Valar Horse Boarding & Transport Phone & Fax: 307-733-2733
Pia Valar   E-mail: valarpia@aol.com
1770 E. River Drive [83001] **Directions:** 1/4 mile from Hoback junction were Hwy 189, 26-89, & 191 meet. 12 miles South of Jackson. Call for Directions. **Facilities:** 6 indoor stalls, 8 25'x35' paddocks, grain/grass/hay/alfalfa available, 1 power and water hookup for trailers with living quarters, 40' round pen, access to national forest trails, custom horse transportation available. (see page 299) **Rates:** $25 per stall, $15 per paddock. **Accommodations:** Hoback River Resort Hotel, Motels & Cabins, and KOA Campground all within a short walk.

LANDER
Lander Sports Arena Phone: 307-332-9790
Douglas Anesi
40 Pheasant Run Drive [82520] **Directions:** Next to airport. Call for directions. **Facilities:** One 50' x 100' pen, seven 20' x 40' pens, no shelters, water available in 2 pens, 240' x 310' outdoor arena, 120' x 300' indoor arena with bleachers & concessions that seats 1,000. Arena used for rodeos, dressage events, & team roping events. No watchman or security. **Rates:** No charge. **Accommodations:** Holiday Lodge & Pronghorn Lodge 1 mile from arena.

Sandstone Ranch Equine Motel LLC Phone: 307-332-2177
Kathryn Kulcher   Fax: 307-335-9535
2529 Sinks Canyon Road [82520] **Directions:** 2.5 miles from Hwy 287. Call for Directions. **Facilities:** 2 12X24, sheds with 24X24, runs, 2 12X12, with 24X12 runs, 1 12X36, shed with roughly 3600, paddock. solar water heater, salt blocks, rubber mats, 70, round pen, feed/hay available, trailer parking, may use Double A Ranch facilities (Indoor/Outdoor Arena) for $10/horse/day. **Rates:** $20 per night. **Accommodations:** Several Hotels within 3 miles.

LARAMIE
*✱ **A. Drummond's Ranch B & B** Phone: 307-634-6042
Taydie Drummond
399 Happy Jack Road (State Hwy 210) [82007]
See listing under CHEYENNE.

Delancey Training Stables Phone: 307-742-2933
Niki & Bernie Delancey
790 Huron [82070] **Directions:** 1/4 mile off I-80 & Hwy 287. Call for directions. **Facilities:** 8-12' x12' box stalls, 20-10' by 20' runs with sheds, 2 large paddocks, automatic waterers throughout. 200' x 60' indoor arena, 50' x 50' indoor round pen, 3 outdoor arenas & a round pen. Trailer parking and custom tack & repair shop on premises. No advance notice needed. **Rates:** $10 per night. Hay extra. **Accommodations:** 5 min. to Holiday Inn, Days Inn & Motel 6.

WYOMING   PAGE 257

ALL OF OUR STABLES REQUIRE CURRENT NEG. COGGINS, CURRENT HEALTH PAPERS, & OWNERSHIP PAPERS.

LARAMIE
On A String Ranch Phone: 307-742-4723
Mernie Younger E-mail: onastring9@aol.com
Web: www.onastring.com
900 Howe Road [82070] **Directions:** 2 miles off of I-80. Call for directions. **Facilities:** 28 indoor 12' x 16' & 12' x 12' stalls, 640 acre pasture, 12 outside paddocks each holding 1-8 horses, 60' x 120' indoor arena, feed/hay & trailer parking available. Horse training available. Call for reservation. **Rates:** $15, including feed, per night. **Accommodations:** Holiday Inn & Motel, and many others nearby.

Stonehouse Stables Phone: 307-742-7512
Don Pratt
3070 Snowy Range Road [82070] **Directions:** 1-3/4 miles off of I-80. Call for directions. **Facilities:** 34 stalls, 10,000 sq. ft. indoor arena, 15,000 sq. ft. outdoor arena, hay & trailer parking available. Training & lessons in driving, reining, cutting, dressage, & Western pleasure. Call for reservation. **Rates:** $20 per night; $100 per week. **Accommodations:** Best Western & Camelot 2 miles.

MEETEETSE
* **Mountain Valley Horse Center** Phone: 307-868-2442
Roxanne & Travis Richardson
P.O. Box 451 [82433] **Directions:** Located 30 miles south of Cody off of Hwy 120. Call for directions. **Facilities:** 6 indoor stalls, 4 outside stalls, large corrals, feed/hay & trailer parking available. Farrier on premises. Trains horses for pleasure. Riding lessons in English & Western. Horses for sale. Call for reservation. **Rates:** $15 per night. **Accommodations:** Motor home that sleeps 6 available for rent. Vision Quest Motel & Oasis Motel 2 miles from stable.

PINEDALE
* **Pole Creek Ranch Bed & Breakfast** Phone: 307-367-4433
Dexter & Carole Smith
244 Pole Creek Road [82941] **Directions:** 1/2 mile south of Pinedale turn up Pole Creek Road and go 2.44 miles, ranch on right. **Facilities:** 5 indoor stalls, 7 acres of pastures, grass/hay available, trailer parking. Horses must be wormed. **Rates:** $5 per night, $15 per week. **Accommodations:** B&B on premises; $55 per night double occupancy. Motels in Pinedale, 3 miles from ranch.

RAWLINS
Carbon County Fairgrounds Phone: 307-324-8101
Spruce & Harshman [82301] **Directions:** 3 blocks off of I-80 & Hwy 287. **Facilities:** 48 indoor stalls, 300' x 300' outdoor arena, lunging & exercise area, no feed/hay, trailer parking available. Rodeos & demolition derby held at grandstand that seats 3,600. **Rates:** $5 per night; ask for weekly rate. **Accommodations:** Rawlins Inn, Days Inn, & Key Motel 1 mile from stable.

WYOMING

ALL OF OUR STABLES REQUIRE CURRENT NEG. COGGINS, CURRENT HEALTH PAPERS, & OWNERSHIP PAPERS.

ROCK SPRINGS
✻ **Rock Springs KOA Campground** Phone: 307-362-3063
Sandy Cochran
86 Foothill Blvd. [82901] **Directions:** Take Exit 99 off of I-80. Follow signs east one mile. Open April-Oct. 15. **Facilities:** 3 outdoor stalls with lean-to for protection, 5-acre fenced sagebrush, feed/hay with prior notice, trailer parking. **Rates:** Horse corral free to persons renting campsite or cabin at facility.

TEN SLEEP
Ten Broek RV Park & Cabins Phone: 307-366-2250
Darell D. Ten Broek E-mail: tenbroekrv@tctwest.net
Box 10, 98 - 2nd Street [82442] **Directions:** 68 miles west of Buffalo on Hwy #16 off I-90. Call for directions. **Facilities:** 10 outdoor stalls, 10 indoor stalls (9' x 9') 6 large corrals, 80' x 150' pasture. 50 x 90 barn and riding area. Feed/hay available, trailer parking with electric, water, sewer, & cable TV. Guided rides available. Fantastic scenery of Big Horn Mountains. Farrier available by advance appointment. Brand ownership required. **Rates:** $5 per night plus feed. **Accommodations:** Valley Motel 3 blocks away; Log Cabin Motel 2 blocks away.

WORLAND
Washakie County Fair Phone: 307-347-8989
602 Fifteen Mile Rd [82401] **Directions:** From US Hwy 16 or US Hwy 20 go to West side of Worland; turn north on Wyoming 433 (West River Road). Fairgrounds 1/4 mile on the left behind cemetery. **Facilities:** 33 permanent stalls, 60 temporary stalls, 14'x14' covered wood stalls. Feed/hay available at local feed stores, trailer parking available. 170'x300' outdoor arena. County Fair in August, USEA recognized horse trails in June, cross-country course nearby, thousand of acres of public land for trail riding. **Rates:** $15 per night plus cleaning deposit. **Accommodations:** Chamber of Commerce (307-347-3226). Six motels in town.

NOTES AND REMINDERS

PAGE 260 **CANADA**

TOWNS SHOWN ARE STABLE LOCATIONS.

* onsite accommodations

CANADA

ALL OF OUR STABLES REQUIRE CURRENT NEG. COGGINS, CURRENT HEALTH PAPERS, & OWNERSHIP PAPERS.

ALBERTA

LETHBRIDGE
✱ Diamond Hitch Acres   Phone: 403-381-4042
Bryan & Tina Smith
Web: diamondhitch. CA   E-mail: tbsmith@telusplanet.net
Box 82 Diamond City (TOK.OTO) **Directions:** 7 miles north of Lethbridge on Hwy 25. **Facilities:** 3 indoor stalls. Feed/hay available. Trailer parking with full electric available. 6 large paddocks. **Rates:** Call for rates. **Accommodations:** Bed, Bales & Breakfast. Log cabin plus rooms in house, just off pavement enroute to Calgary, Edmonton and Alaska.

NORTH COOKING LAKE
Still Meadows Ranch   Phone: 780-922-5566
Rusnak Family   or: 780-990-7229
Range Road 210A [T8G1G6]
Directions: Call for directions. 2-1/4 kilometers north of Hwy 14.
Facilities: 32 indoor stalls, 70' x 200' indoor arena, 2 outdoor arenas, outside pens, feed/hay, trailer parking. 160 acres with trails. Paints & quarter horses on premises. Standing-at-stud: "Chattanooga Choo Choo." **Rates:** $10 per night. **Accommodations:** Tofield Campground within 10 minutes.

BRITISH COLUMBIA

POUCE COUPE (DAWSON CREEK area)
✱ Red Roof Bed & Breakfast   Phone: 250-786-5581
Laurie Embree
Box 727 [V0C 2C0] **Directions:** On Blockline Road, 3 miles west of Hwy 2. Call for further directions. **Facilities:** 1 log 25' round corral. 1-10 acre pasture/turnout with barb wire fences, feed/hay, trailer parking available. **Rates:** $5 per night. **Accommodations:** B&B on premises.

ALL OF OUR STABLES REQUIRE CURRENT NEG. COGGINS, CURRENT HEALTH PAPERS, & OWNERSHIP PAPERS.

MANITOBA

WINNIPEG
Poco-Razz Farm Phone: 204-255-4717
Jim Shapiro & Christina Eyres
130 Greenview Road [R3V 1L6] **Directions:** From Perimeter Hwy (#100) go 5 kilometers or 3 miles south on St. Mary's Road. Call for directions. **Facilities:** 2 indoor 9' x 12' stalls, 2 indoor 10'x12' stalls, 12' aisleway; 3 large corrals, one of which is 90' x 110' outdoor arena, 2 acres of pasture, oat straw bedding, water available in each stall and in corrals, hay available, no barbed wire, rubber fencing with hot wire; wash area & hot water tank, muck-out area, parking for horse trailers. Vet nearby; farrier available. Trails for riding, fishing. **Rates:** $30 per night outdoor. **Accommodations:** Comfort Inn 4.2 miles from farm in Winnipeg.

NEW BRUNSWICK

HARVEY STATION
Holiday Ranch Phone: 506-366-3291
Gary & Brenda Nason
Rte. 636 [E0H 1H0] **Directions:** Call for directions. About 4 miles off Hwy 3; 15 miles from Trans-Canada Hwy. **Facilities:** 6 box stalls, 40' x 60' indoor arena, large outdoor arena, feed/hay, trailer parking. Clinics available. Crown ground behind ranch with endless trails. Call for reservations. **Rates:** $10 per night. **Accommodations:** B&B in area. Motel in Fredricton 20 miles away.

PETITCODIAC
Sheffield Stables Phone: 506-756-1110
Charles & Paula Jacob
RR #5 [E0A 2H0] **Directions:** Call for directions. Right off Petitcodiac exit of Trans-Canada Hwy. **Facilities:** 35 indoor 10' x 12' stalls, 250' x 150' and 150' x 150' outdoor arenas, 50' x 100' indoor arena, 9 separate paddocks, feed/hay, trailer parking available. Miles of trails, camp on trails, mechanic on property, carriage & sleigh rides. Instruction & training, 2 CEF shows a year, cross-country course. Call for reservation. **Rates:** $25 per night. **Accommodations:** Motels in Moncton, 15 miles away.

CANADA

ALL OF OUR STABLES REQUIRE CURRENT NEG. COGGINS, CURRENT HEALTH PAPERS, & OWNERSHIP PAPERS.

NOVA SCOTIA

GRANVILLE FERRY
Equus Centre　　　　　　　　　　　　　　Phone: 902-532-2460
Jennifer Gale
Box 1160, RR #1, 5613 [B0S 1K0] **Directions:** Call for directions. Located 5 minutes from historic Annapolis Royal. **Facilities:** 23 indoor stalls most 11' x 12', pasture/turnout, 60' x 136' indoor arena, hunt course, jumping ring, 20' x 60' dressage ring, feed/hay, trailer parking. Trails, lessons. Call for availability. **Rates:** $6 per night. **Accommodations:** Motels and B&Bs within 5-10 minutes.

MILTON
Birch Lane Farm　　　　　　　　　　　　Phone: 902-368-1113
John McAssey
Rte. 2 [C1E 1Z2] **Directions:** Call for directions. 7 kilometers west of Charlottetown. **Facilities:** 33 indoor stalls 8' x 10' to 12' x 10', turnout, 60' x 120' indoor arena, outdoor riding ring, tack room, feed/hay, trailer parking. Heated classroom, viewing lounge, lessons, training. Call for reservation. **Rates:** $10 per night. **Accommodations:** Motels in Winslow (2 kilometers) and Charlottetown (7 kilometers).

MOUNT UNIACKE
Atlantic Dressage Development Center　　　Phone: 902-866-1198
Debbie Trimper
Old Windsor Hwy [B0N 1Z0] **Directions:** Call for directions. 25 minutes out of Halifax. Off Exit 3 on Hwy 101. **Facilities:** 37 indoor stalls, paddocks, indoor & outdoor arenas, 1/2-mile track, feed/hay, trailer parking. Competition stable, riding lessons. Call for availability. **Rates:** $20 per night. **Accommodations:** Motels in Sackville, 10 miles.

ONTARIO

BADEN
* **North Ridge Farm**　　　　　　　　　　　Phone: 519-634-8595
Sarah Banbury
RR #21 [N0B 1G0] **Directions:** Call for directions. Between Stratford, Kitchener, Waterloo. 6 kilometers from Hwy 7/8, 20 minutes north of Hwy 401. **Facilities:** 6 indoor stalls 10' x 12' or larger, limited pasture/turnout, feed/hay, trailer parking. Complete equestrian facility, outdoor arena, indoor ring, lessons. Call for reservations. **Rates:** $20 per night includes hay and bedding. **Accommodations:** Bed & breakfast on premises, $45 single, $60 double.

ALL OF OUR STABLES REQUIRE CURRENT NEG. COGGINS, CURRENT HEALTH PAPERS, & OWNERSHIP PAPERS.

ONTARIO

BOWMANVILLE
Colonial Equestrian Centre   Phone: 905-623-7336
Mrs. T. Ashton
3706 Rundle Road, RR #3 [L1C 3K4] **Directions:** Call for directions. Located close to Hwy 2 & Hwy 401. **Facilities:** 55 indoor 10' x 12' stalls, pasture/turnout, 60' x 12' indoor arena with heated viewing lounge, 3 large barns, 12 large paddocks, heated tack room, feed/hay, trailer parking. 27 acres for riding, adjoining wooded trails, hunter/jumper courses, dressage area, all wood fencing. **Rates:** $25 per night. **Accommodations:** Motels within 4 kilometers of Centre.

PORTLAND
✣ Box Arrow Farm   Phone: 613-272-2882
Ruth Godwin or Shirley Prosser   or (Summer only): 613-272-2509
Cove Road, Big Rideau Lake [K0G 1V0] **Directions:** If you are coming from Kingston, turn on Hwy 15. If you are coming from Montreal, turn on Hwy 32 and it runs into Hwy 15. **Facilities:** 60 indoor 12' x 12' stalls, 6 good-sized paddocks, hay available but bring own feed, trailer parking. Lots of trails, event course. October Poker Run, Holiday Horse related place. Call for reservations. Cottages on premises, swimming, fishing. **Rates:** $25 per night; $125 per week. **Accommodations:** Cottages on the water on premises, $100 per day, $500 per week. Many lodges and small inns within 15 miles plus 4 golf courses.

ROCKWOOD
Travis Hall Equestrian Centre   Phone: 519-843-4293
Dave and Judith Johnson   Fax: 519-843-4903
RR #3 [N0B 2K0]   Web: www.geocities.com/travishallequestrian
Directions. 6 kilometers north of Hwy 24, up to Wellington 29, turn off Eramosa 30th, first farm on right. **Facilities:** 35 stalls, pasture for turnout, feed/hay, trailer parking, RV hook-ups available. Miles of gravel road for trail riding, cross-country course. Specialize in English, Western, and driving. Call for reservations. **Rates:** $20 per night. **Accommodations:** Motels in Guelph and Fergus, 15 minutes from stable.

ZEPHYR
✣ High Fields Ranch/Country Inn & Spa   Phone: 905-473-6132
Norma Daniel   Fax: 905-473-1044
11568-70 Concession 3 (Mail: P.O. Box 218) [L0E 1T0] **Directions:** Don Valley Parkway north, Hwy 404 North, east on Davis Drive (last exit) for 16.5 kilometers, 7.5 kilometers north on Concession 3, west side. **Facilities:** 19 indoor 10' x 12' stalls, 2-acre pasture, hay, trailer parking. **Rates:** $20 per night. **Accommodations:** Country Inn & Spa on premises.

ALL OF OUR STABLES REQUIRE CURRENT NEG. COGGINS, CURRENT HEALTH PAPERS, & OWNERSHIP PAPERS.

SASKATCHEWAN

PRINCE ALBERT
<u>Asiil Arabians</u>　　　　　　　　　　　　　　　Phone: 306-764-7900
Jack & Harriet Lang
Box 275 SK [S6V 5R5] Directions: Call for directions. 8 miles southeast of town off Hwy 3. **Facilities:** 8 indoor large box stalls, pasture/turnout, outdoor riding arena, feed/hay, trailer parking. Trails close by; good riding within 10 minutes. Straight Egyptian breeding operation. Call for reservation. **Rates:** $5 per night. **Accommodations:** Motels within 5 minutes.

<u>Red River Equestrian Center</u>　　　　　　　Phone: 306-763-3434
Gord Trueman
Box 10, Site 19, RR #5 [S6V 5R3] Directions: Call for directions. 3 kilometers north on Hwy 2, east at Whispering Pines Trailer Court, facility 1/2 mile on south side. **Facilities:** 47 indoor 10' x 12' box stalls and 5' x 8' tie stalls, turnout available, feed/hay, trailer parking. Nightly lessons; roping, cutting, English and Western riding, jumping. Sleigh rides and wagon rides. Call for availability. **Rates:** $10 per night. **Accommodations:** Motels within 4-5 kilometers.

REGINA
<u>Twin Pine Stables</u>　　　　　　　　　　　　Phone: 306-757-4882
Corace & Chris Pedersen
Box 405, Off Hwy 1 [S4P 3A2] Directions: Call for directions. Off Hwy 1 west of Regina. **Facilities:** 20 indoor box stalls 8' x 12' and 8' x 10', pasture/turnout, large indoor riding arena, heated barn, feed/hay, trailer parking. Riding lessons. **Rates:** $10-$20 per night depending on services. **Accommodations:** Motels within 3 kilometers in Regina.

NOTES AND REMINDERS

NOTES AND REMINDERS

courtesy of
USRider

TRAILERING TIPS

Veterinarian -Approved First Aid Kit

Carrying the following items in an emergency kit will help you handle most situations. Discuss this list with your own veterinarian; he/she may have other suggestions appropriate for you and your situation.

ROLL COTTON – 2 rolls
ROLL GAUZE – 4 rolls
GAUZE SQUARES
CLEAN STANDING BANDAGES
2 quilt or fleece with outer wraps
ADHESIVE TAPE
24" SECTION OF 6" PVC PIPE
which has been split in half length-wise
for splinting: check that diameter of pipe fits your horse.
COHESIVE FLEXIBLE BANDAGE
2 Vetratp® – Co-flex®
STICKY ROLL BANDAGE
Elastikon®
THERMOMETER
STETHOSCOPE
MOSQUITO FORCEPS
SCISSORS
TWITCH
ANTISEPTIC SOAP
Betadine®
HYDROGEN PEROXIDE
ANTIBACTERIAL OINTMENT
ANTIBACTERIAL SPRAY POWDER
OPHTHALMIC OINTMENT
SALINE EYE WASH
BUTAZOLIDIN PASTE
BANAMINE GRANULES OR PASTE
BUCKET
WATER
10 gallons or more

All medications should be given at the advice of your veterinarian or the veterinarian treating the condition.

IMPROPER USE OF TRANQUILIZERS AND OTHER MEDICATIONS CAN RESULT IN THE LOSS OF YOUR HORSE.

Equine Trailering Tips courtesy of USRider Equestrian Motor Plan in cooperation with nationally respected equine trailering expert and author Neva Kittrell Scheve and veterinarian and author James M. Hamilton, DVM. For more trailering information, visit the Equine Trailer Safety Area at www.usrider.org.

www.usrider.org

VALAR

Horse Boarding & Transport
Your Custom Care Specialists

Pia Valar
1770 E. River Drive, Hoback Junction
P.O. Box 3365, Jackson, WY 83001
E-mail: valarpia@aol.com

307-733-2733

GRAPHICS PLUS

Freelance Graphic Design

Logo Design • Ads • Illustration
Corporate Identity • Brochures
Posters • Newsletters • Books
Forms • Desktop Publishing • Etc.

Tel: 781-934-2818
E-mail: cherry8@adelpha.net

Horse Transportation

We Care About Your Animals

Experienced Horsemen

Coast To Coast Twice Monthly

HORSEIN' AROUND

1-800-234-4675

TRAILERING TIPS

Fuel Efficiency

What can horse owners do to improve a truck's efficiency in these days of rising gas prices?

Vehicle Maintenance:

    Keep Engine Properly Tuned – Depending upon the kind of repair done, this can result in an average 4 percent increase in fuel efficiency. Replacing a faulty oxygen sensor can improve fuel mileage as much as 40 percent.

    Check & Replace Air Filter – Replacing a clogged air filter can improve your vehicle's mileage up to 10 percent.

    Keep Tires Properly Inflated – Proper inflation can increase your mileage by around 3 percent. Added benefit: Properly inflated tires are safer and last longer.

    Use Recommended Grade of Motor Oil – Using the incorrect weight can increase fuel consumption by 1-2 percent. Look for motor oil that says "Energy Conserving" on the API performance symbol to be sure it contains friction-reducing additives.

Vehicle Operation:

    Drive Sensibly – Aggressive driving can lower your gas mileage by one-third.

    Observe the Speed Limit – Each 5 mph you drive over 60 mph is like paying an additional 10 cents per gallon of gas. Added benefit: Observing the speed limit is also safer.

    Avoid Excessive Idling – Idling gets 0 miles per gallon.

    Use Cruise Control – Using cruise control (where applicable) helps you maintain a constant speed and, in most cases, will save gas.

    Use Overdrive Gears – When your engine speed goes down, your mileage goes up. Added benefit: Using overdrive gears reduces engine wear.

Equine Trailering Tips courtesy of USRider Equestrian Motor Plan in cooperation with nationally respected equine trailering expert and author Neva Kittrell Scheve and veterinarian and author James M. Hamilton, DVM. For more trailering information, visit the Equine Trailer Safety Area at www.usrider.org.

www.usrider.org

courtesy of

Trailering Tips

Storage Preparation

If your trailer is to be stored for an extended period or over the winter, it is important that the trailer be prepared properly.

1. Remove the emergency breakaway battery and store inside, out of the weather. Charge the battery at least every 90 days.

2. Jack up the trailer and place jack stands under the trailer frame so that the weight will be off the tires. Follow trailer manufacturer's guidelines to lift and support the unit. Never jack up or place jack stands on the axle tube or on the equalizers.

3. Lubricate mechanical moving parts, such as the hitch and suspension parts that are exposed to the weather.

Note: On oil-lubricated hubs, the upper part of the roller bearings are not immersed in oil and are subjected to potential corrosion. For maximum bearing life it is recommended that you revolve the wheels periodically (every 2 to 3 weeks) during periods of prolonged storage.

Equine Trailering Tips courtesy of USRider Equestrian Motor Plan in cooperation with nationally respected equine trailering expert and author Neva Kittrell Scheve and veterinarian and author James M. Hamilton, DVM. For more trailering information, visit the Equine Trailer Safety Area at www.usrider.org.

www.usrider.org

courtesy of
USRider

TRAILERING TIPS

The following list of additional suggestions should help make trailering easier for you and your horse:

- Wear gloves and boots when you are loading and unloading.

- If the trailer is dark inside when you are loading, open the doors and turn on the lights to make it lighter.

- If you are having trouble loading a horse, at least ten well-meaning bystanders will show up to help you. Thank them for offering to help but ask them all to leave except those who you know will be able to help. Too many cooks in the kitchen can really make a situation worse.

- Make sure there are no hazards near the trailer when you are loading and unloading, farm machinery, fence posts, etc.

- Don't let door covers stick out the sides where a horse handler could get bumped in the head.

- If two or more horses are being unloaded from the trailer, keep at least one horse in sight of the last horse until he has also been safely unloaded. The one that is left on the trailer may panic and rush off too quickly. This is more of a problem with an inexperienced horse.

- If you are hauling your horse in someone else's trailer, do your own safety check. Don't depend on someone else for your safety and the safety of your horse.

- If you are hauling someone else's horse in your trailer, insist the horse wear protective bandages, and agree in advance who will be responsible in the event of injury to the horse or damage to the trailer. Check with the insurance company to see who is covered for what.

- Don't travel alone if you can help it.

- Never lead a horse into the trailer if you do not have an escape route.

- Never get into a trailer with a panicked horse, and don't open the door if there is a chance the horse could bolt out the door onto the highway.

- Never put a horse into a trailer that is unhitched, or unhitch a trailer while the horse is in it.

- Don't use tranquilizers unless you know how. Improper use of tranquilizers can cause death.

- Discuss the use of tranquilizers with your veterinarian.

Equine Trailering Tips courtesy of USRider Equestrian Motor Plan in cooperation with nationally respected equine trailering expert and author Neva Kittrell Scheve and veterinarian and author James M. Hamilton, DVM. For more trailering information, visit the Equine Trailer Safety Area at www.usrider.org.

www.usrider.org

courtesy of
USRIDER

TRAILERING TIPS

Trip Preparation: Last-Minute Checklist

- Check the tow vehicle. Check and replenish engine fluid levels and wiper fluid. Towing puts extra stress on the radiator, brakes and transmission. Make sure fluid levels are correct.

- Make sure the ball on the tow vehicle is the correct size for the trailer.

- Make sure the rearview mirrors are properly adjusted and you know how to use them.

- Check tires pressure in the tires of the tow vehicle and the trailer. Improper tire pressure is responsible for most towing problems. Check tire condition.

- Make sure that the trailer is level so the horse is not always fighting his balance by traveling uphill or downhill. This movement can also cause the trailer to sway and cause other safety problems.

- Check lug nuts on wheels. Wheel nuts and bolts should be torqued before first road use after each wheel removal. Check and re-torque after the first 10 miles, 25 miles and again at 50 miles. Check periodically thereafter.

- Check inside the trailer for bee and wasp nests.

- Check over your hitch, coupler, breakaway brake battery and safety chains. Make sure the brakes and all lights are working properly before you load the horses.

- When horses are loaded, make sure all doors are latched properly and horses are tied.

- Drive down the driveway, and before you drive onto a main road, get out and check over everything again. Something you overlooked may make itself apparent by then. Most accidents happen to people who have been hauling just long enough to get lackadaisical.

- If you happen to stop somewhere where the rig has been left attended, check everything all over again. Someone may have been tampering with the trailer or horses.

Equine Trailering Tips courtesy of USRider Equestrian Motor Plan in cooperation with nationally respected equine trailering expert and author Neva Kittrell Scheve and veterinarian and author James M. Hamilton, DVM. For more trailering information, visit the Equine Trailer Safety Area at www.usrider.org.

www.usrider.org

courtesy of
USRider

TRAILERING TIPS

Neva Kittrell Scheve is author of **Hawkins Guide: Horse Trailering on the Road** and, **The Complete Guide to Buying, Maintaining, and Servicing a Horse Trailer** and co-author of **Hawkins Guide: Equine Emergencies on the Road**. She travels nationwide to give seminars on all aspects of horse trailers. With her husband, Thomas G. Scheve, she has developed numerous lines of horse trailers, which are marketed internationally through their own company, **EquiSpirit**, located in Southern Pines, NC. Neva is has been a horsewoman for over 30 years and competes in dressage and carriage driving. She is also a member of the Moore County Equine Emergency Response Unit in NC, and the VMAT (Veterinary Medical Assistance Team) which is a part of FEMA (Federal Emergency Management Agency).

For more information about EquiSpirit Trailers or to purchase their indispensable equine travel publications, call **toll-free 1-877-575-1771**. to request a free brochure or visit **www.equispirit.com**. **USRider** members receive a discount on their publications. For more information about USRider, call **toll-free 1-800-844-1409** to request a free brochure or visit **www.usrider.org**.

www.usrider.org

courtesy of

Trailering Tips

James M. Hamilton, DVM, originally from Cincinnati, Ohio, Jim rode horses and was involved in Pony Club and 4-H before earning his bachelor's degree and doing post-graduate studies at Ohio State. After graduating from the University of Georgia School of Veterinary Medicine, he joined a sports medicine practice at Belmont Park in New York.

His interest is in equine sports medicine, especially respiratory, orthopedics and imaging diagnostics. He now resides and practices in the Sandhills as a partner in **Southern Pines Equine Associates**. The practice provides complete medical and surgery services to the rapidly growing horse community in southern North Carolina. Dr. Hamilton is the co-author of **Equine Emergencies on the Road**, a glove compartment manual on prevention and treatment of en route illness and injury of horses. This manual was on the United States Pony Club's "must read" list for 1995.

In 1997 he became involved in the AVMA's Disaster Response Program as Team Commander of VMAT-3. This team of veterinarians and technicians is now part of the federal government's National Disaster Medical System, a department within the United States Public health Health Service (USPHS). Most recently VMAT#3 responded along with others to the World Trade Center disaster in NYC. Team members were there to support the federal government's (FEMA) Search and Rescue dog teams. Dr. Hamilton is a member of the American Veterinary Medical Association, American Association of Equine Practitioners, American Academy of Veterinary Disaster Medicine, and past member of the United States Pony Club's National Safety Committee. In 1998, Jim earned the **"Veterinarian of the Year"** award for the work he did to help the state veterinary association (NCVMA) organize a disaster plan.

He is currently serving on a statewide committee **(SART)** working with State Emergency Management officials, counties, industry representatives and others to create a comprehensive animal disaster response plan for North Carolina.

www.usrider.org

NOTES AND REMINDERS

About the Publisher

Lisa Doubleday began her love affair with horses as a young girl growing up in the Northeast. For the past several years Lisa was editor of the **US Stabling Guide** and vice president of Custom Ranch Vacations and published the official DRA (Dude Ranchers' Association) **Custom Ranch Vacation** book. Her close relationship with the members of the DRA has the added benefit of spending time on different ranches and reinforces her love of the west. In 2003 Lisa founded Lariat Publications and now publishes the **US Stabling Guide**. Lisa owns her dream horse, a palomino Quarter Horse named Premo. She enjoys trail riding, cutting, fly fishing and pack trips. Lisa has a Masters Degree in Education, and hosts clinics for people with fear of horses. Lisa currently resides in Massachusetts.

NOTES AND REMINDERS

NOTES AND REMINDERS

NOTES AND REMINDERS

NOTES AND REMINDERS

NOTES AND REMINDERS

NOTES AND REMINDERS

NOTES AND REMINDERS

NOTES AND REMINDERS

NOTES AND REMINDERS

NOTES AND REMINDERS

VACATION PLANNING?
FREE DIRECTORY.

THE DUDE RANCHERS' ASSOCIATION
P.O. BOX 2307, CODY, WY · 307-587-2339
WWW.DUDERANCH.ORG · INFO@DUDERANCH.ORG